DATE DUE

OCT 0 4 2007	
OCT 0 6	

BRODART, CO.   Cat. No. 23-221-003

The Smithsonian Book of

Smithsonian Library of the Solar System

Edited by Ted A. Maxwell

This series is intended for the general audience but finds equal use by specialists in the planetary sciences. Written by scientists who are experts in their fields, these volumes present the story of the missions and the results of recent exploration and review the latest theories of the origin of our neighbors in the solar system.

Also in the Series:

The Once and Future Moon
Paul D. Spudis

Jupiter: The Giant Planet
Second Edition
Reta Beebe

Mercury: The Elusive Planet
Robert G. Strom

The Restless Sun
Donat G. Wentzel

The Smithsonian Book of

Mars

Joseph M. Boyce

Smithsonian Institution Press
Washington and London

Copy Editor: Eileen D'Araujo
Production Editor: Ruth G. Thomson

Library of Congress Cataloging-in-Publication Data
Boyce, Joseph.
   Smithsonian book of Mars / Joseph M. Boyce.
      p.   cm.
   Includes bibliographical references and index.
   ISBN 1-58834-074-0
      1. Mars (Planet). I. Title.
   QB641.B68   2002
   523.43—dc21   2002018832

British Library Cataloging-in-Publication Data available

Manufactured in China
09 08 07 06 05 04 03 02 5 4 3 2 1

∞ The paper used in this publication meets the minimum requirements of the American National Standard for Information Sciences—Permanence of Paper for Printed Library Materials ANSI Z39.48-1984.

This book is dedicated to my friends and mentors:
Mike, Larry, David, Chuck, and Steve

Contents

Foreword

Mars has fascinated humanity for centuries and continues to dazzle even after nearly forty years of space exploration. Telescopic charting of the Martian surface paved the way for an initial wave of seemingly primitive flybys during the first stage of spaceborne reconnaissance of the Red Planet. These initial forays provided tantalizing views and essential data pertaining to the environmental conditions that describe the current Martian climate. Indeed, the initial Mariner flybys of the 1960s by the National Aeronautics and Space Administration (NASA) dispelled many misconceptions about Mars and provided a humbling revolution in scientific thinking that is being challenged only today, thanks to a new class of observations.

We learned that Mars was far from the hospitable world imagined by the science fiction writers and was best viewed as a vast polar desert, dominated by the action of wind-borne dust in its modern era and seemingly "lunar" in its long-distant past.

Just as Renaissance exploration of planet Earth forever changed humankind's view of the home planet, the orbital and lander-based exploration missions of the 1970s, starting with Mariner 9 and continuing with Viking, forever altered the modern view of Mars. Mariner 9 instantly shattered the "lunar-Mars" concept by revealing the full tapestry of Martian landscapes, including gigantic canyonlands and shield volcanoes, as well as polar ice caps and impact basins. Furthermore, Mariner 9 illustrated the many possible roles of liquid water in the Martian "system" by discovering evidence of trellislike valley networks and massive outflow channels, harbingers of an ancient warmer and wetter Mars. Viking followed Mariner 9 with an almost unimaginable array of orbital and landed instruments; it was indeed a "robotic Apollo mission" that attacked the ongoing mysteries of whether Mars harbors or ever harbored a biosphere. Although

Viking failed to identify compelling evidence of an extant biosphere, the emerging view of Mars painted by the rich array of Viking observations was that of a planet with a dynamic past.

It was perhaps much wetter and warmer than the inhospitable present—a planet relatively dormant now, with little possibility of an accessible biological record. With this new view in mind, future exploration of Mars was directed to a continuation of the global reconnaissance needed to fill in gaps and to set the stage for basic in situ characterization.

The nearly unimaginable loss of Mars Observer in 1993 was a setback. But, thanks to the heroic efforts of a team of engineers and scientists at NASA's Jet Propulsion Laboratory, the successful landing of Mars Pathfinder and demonstration of its Sojourner Rover renewed interest in Martian exploration. In late summer of 1996, a team of researchers published a highly contentious article in the journal *Science* suggesting that submicrometer-scale microbes were preserved in a meteorite that was believed to have arrived on Earth from Mars. Although the evidence was highly contested, additional observations of other possibly Martian meteorites also pointed to hydrocarbons in regular patterns at nano-scales, reawakening interest in the possibility of life on Mars. These findings catalyzed the development of an "astrobiology initiative" within NASA that capitalized on the emerging biological revolution on Earth and applied it to the exploration of the Universe.

Central to the overarching question that is one of NASA's strategic objectives is "are we alone," and Mars offers an initial proving ground for exploring this poignant issue. While the scientific community struggled with the many issues associated with the possibility of life preserved within Martian meteorites, NASA orbited the intrepid Mars Global Surveyor spacecraft, starting in 1997. By early 2001 Mars Global Surveyor had mapped the entire planet in new wavelengths, at unprecedented scales, and provided a new context for exploring Mars. Mars Global Surveyor literally put a new face on Mars, overturning theories that had been staunchly defended for 25 years, with dramatic and definitive evidence of layers, modern emergence of fluids, climate variability, an ancient but long-gone magnetic field, and a history of lithospheric evolution that involved impact, tectonism, water, and volcanism. The new Mars that is currently emerging offers a scientific smorgasbord for virtually all disciplines of space science, ranging from astrobiology to the physics of neutrons.

Today we are poised to unravel yet more of the previously unseen Mars with the arrival of Mars Odyssey Orbiter and its array of remote sensing instruments. What will be found is unknown, but prospects are many. It is possible that Odyssey will unearth Martian "hydrogen hot spots" with its neutron spectrometer, discover evidence of mineralogical "oases" where liquid water resulted in local surface exposures of diagenetic minerals, and perhaps even find evidence of festering Martian "hot spots."

Beyond Odyssey, NASA will return to the Martian surface as early as 2004 to explore two compelling sites for local evidence of water-related materials and to assess the planet mineralogically at hand-lens scales. These near-term missions will give way to an unprecedented wave of both orbital and surface-based reconnaissance, culminating within a decade or so in the first robotic mission to return samples from another planet. The initial Mars Sample Return mission will likely be the first of several, and ultimately the time will come when human explorers may serve as the agents of sample return, just as they so capably performed on the Apollo voyages to the Moon in the late 1960s and early 1970s.

As we ponder the many pathways that the future of Mars exploration will bring, there remains an imperative to bring the cognitive and field-adaptability powers of the human explorer to bear on the many challenging problems that remain. If Mars does truly harbor a preserved biological record, discovering the local field relationships that will ultimately yield such secrets on a planet with a surface area equivalent to the land area of planet Earth may require human exploration at some point. For the moment, however, robotic forerunners of distant human explorers are vital to developing critical answers about the uniquely Martian dynamic systems that dominate the Red Planet.

The first decade of the new millennium at hand could well be remembered as the "decade of Martian discovery" if NASA's restructured Mars Exploration Program delivers the powerful scientific results and datasets that have been promised. Whatever the outcome of the next decade, it is safe to say that Mars will continue to dazzle even the naysayers with the discoveries that are likely to emerge. By 2010 humanity may well be wondering why we are here and not on Mars!

As we look at Mars from inside out in the pages that follow, bear in mind that a scientific revolution is at hand, and within the next 10–15 years much of what we think we know and understand could be supplanted with a new perspective on the never-disappointing Red Planet.

James B. Garvin

NASA Lead Scientist for Mars Exploration
Office of Space Science
NASA Headquarters

Acknowledgments

This book was written over many years, slowed by the time commitments of my involvement in NASA's Mars Program. I was fortunate enough to be living some of the history outlined in this book. As with any such book, there are many people that have helped to produce this volume. I am particularly grateful to Geoffrey Briggs, the former director of NASA's Solar System Exploration Division (and my former boss). When asked by Robert Wolf, the first science editor of the Smithsonian's series on the planets, to write this book, I initially declined. In spite of my reluctance, Dr. Briggs was forcefully persuasive in convincing me to reconsider Dr. Wolf's request.

In addition, I thank Nadine Barlow and Ted Maxwell for their thoughtful reviews of the manuscript. Their considerable efforts have greatly improved the final product. Discussions with such Mars scientists as Michael Carr, Laurence Soderblom, Hal Masursky, Ronald Greeley, James Head, Raymond Arvidson, Bruce Murray, Bruce Jakosky, Chris McKay, James Garvin, Hap McSween, Donald Bogard, David McKay, Gerald Soffen, Michael Meyer, Nadine Barlow, Gearld Schaber, Robert Strom, Victor Baker, Matthew Golombek, Steve Squyres, Joseph Veverka, Gerhard Neukum, William Hartmann, Michael Malin, Maria Zuber, Sean Solomon, Roger Phillips, Peter Mouginis-Mark, Phil Christensen, Maria Acuna, Dave Smith, and Richard Zurek (and many others) have helped me to know Mars better. I am also greatly indebted to Linda Schiender, Peter Doms, Leslie Pieri, Gerald Schaber, Sandra Schaber, Trent Hare, Virginia Hall, Annie Howington, Sharon Fontana, Jeffrey Plescia, and Sandra Dueck for their help and support throughout this project. A special thanks to my daughters, Amy and Heather, who shared my time with the preparation of this manuscript.

I would like to acknowledge the engineers and scientists at the Jet Propulsion Laboratory who have made possible most of the successful missions to Mars and as a result this book. I have always been and continue to be in awe of their wizardry. Finally, although this book was not directly supported by NASA, it was NASA, through my job, that provided me with a front-row seat for one of the most exciting endeavors in human history—the exploration of Mars.

Chapter 1

An Introduction to Mars

More than any other planet in our solar system, Mars has fascinated and intrigued us. From the beginning, when ancient observers first noticed its distinctive red color, continuous movement across the heavens, and peculiar periodic disappearance and reappearance each year, they formed legends and myths to explain these oddities.

As more and more powerful telescopes were invented and turned toward Mars, astronomers began to bring the planet into sharper focus (Figure 1). They could see seasonally changing surface bright and dark markings, as well as seasonal changes in its polar ice caps. Astronomers spied white clouds and dust storms that swept over the surface. The conclusion drawn from these discoveries was simple—the surface was similar to the deserts of Earth. Armed with only this information and a vivid imagination, one of these astronomers, Percival Lowell, single-handedly transformed Mars into the home of a dying race of desert dwellers. His eloquent writings still influence the popular image of Mars, as evidenced by recent movies such as *Mars Attacks!*

Starting in the 1960s, our vicarious exploration of Mars using robotic spacecraft has dramatically changed our perception of the planet. These changes have been systematic. Early missions measured the atmosphere and found it to be thin, cold, dry, and composed mainly of carbon dioxide. Early missions could only see large-scale features on the surface and gave little information about the detailed history of the planet. From these early observations, we began to think that Mars was not much more than "Earth's Moon with a thin atmosphere" (as Jack McCauley of the U.S. Geological Survey described it during a Mariner 9 team meeting). Viking

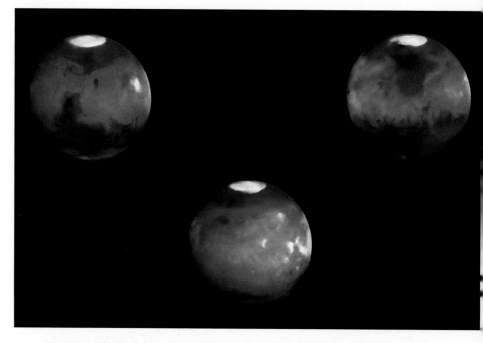

Figure 1. These three views of Mars show the classical markings on the surface of Mars that have been observed since the invention of powerful telescopes. The pictures are some of the highest resolution pictures ever obtained of Mars from Earth. They were taken from the Hubble Space Telescope, when Earth was within about 103 million km (62 million miles) of Mars. The rotation of Mars provided these three views. The picture on the upper left is centered near Syrtis Major, and below is the giant impact basin Hellas. White clouds cover the great volcanoes in the Elysium region. The picture in the bottom center shows a crescent-shaped cloud just right of center hanging over the huge volcano Olympus Mons. These clouds, like those over Elysium, are produced as warm afternoon air pushes up and over the summits, forming ice-crystal clouds downwind of the volcanoes. The picture in the upper right is centered near the Chryse Basin where the Viking 1 lander and Mars Pathfinder spacecraft landed. Valles Marineris, an immense canyon system the length of the continental United States, is seen in the lower left. (Courtesy NASA/Hubble Space Telescope Science Institute)

softened this view only slightly and hammered home that it was unlikely that complex organisms could live on the surface of Mars.

Our impression of what Mars is and what it was has changed systematically with the increasing resolution and the collection of new
types of observations. Recently, Mars Global Surveyor carried a payload of scientific instruments that provided the roots of a revolution
in thinking. It transformed our ideas about Mars as being a place
that has always been a cold, dry, barren desert to a place where surface conditions are periodically mild enough to produce thick sedimentary deposits, a place where early in its history it may have
rained, a place where water still periodically emerges from underground to carve channels, a place where for its first several hundred
million years a global magnetic field shielded its surface from the
killing radiation of the Sun, enabling a condition where life could
have started on the surface.

Although heavily weighted to the recent discoveries made by
Mars Global Surveyor, this book is intended to provide a general
view of Mars built on several hundred years of observations from
ground telescopes and the more than four decades of exploration by
spaceflight missions. As with Viking and the Mariner missions before it, the planet unveiled by Mars Global Surveyor will not be the
final story. Future books will chronicle new discoveries about Mars
that will be every bit as surprising as the discoveries of each past
Mars mission.

Why Read This Book?

My aim here is to provide a broad overview of what is currently
known about Mars: what Mars is now, how it got that way, and how
it might change in the future. This book is intended for everyone
with an interest in Mars. For this reason, I have attempted to keep
the technical jargon and details to a minimum. The book begins
with an introductory overview of what is known about Mars, intended to provide a framework for the detailed discussion that follows. Next is an outline of the evolution of our understanding of
Mars, tracing it from the early stargazers through ground-based telescopic observations (chapter 2) to the exploration of Mars with
spacecraft (chapter 3). This history leads into a discussion of the
planet as we currently know it, beginning with the Martian interior
(chapter 4) and working outward to the surface (chapter 5) and at-

mosphere (chapter 6). Following are chapters on what we know about the search for life on Mars and the potential for finding it (chapter 7) and on its two asteroid-like satellites (chapter 8). The final chapter focuses on future plans for the exploration of Mars. Detailed exploration plans are notoriously dynamic. This final chapter should be read in that context.

Before we move into the main body of the book, we will take a step back and take a general view of the planet. This is intended to provide a foundation and a context for each of the chapters that follow.

The Planet Mars

Mars is the fourth planet from the sun and is the outermost of the terrestrial planets (rocky inner planets) of our solar system (Table 1). It is the second closest planet to Earth after Venus. Mars is 6,778 km (4,067 miles) in diameter, intermediate in size between Earth and Mercury. The surface area of Mars is nearly equivalent in size to the land area of Earth. Mars is made from roughly the same ingredients as Earth: silicates and oxides of iron and magnesium, as well as metallic iron alloyed with other constituents. It is distinctly reddish, caused by the oxidation of the abundant iron in its soils (i.e., the iron has rusted). Mars has a thin, dry atmosphere rich in carbon dioxide that freely exchanges with volatiles contained in its surface materials.

The Motion of Mars

Even before the invention of the telescope, ancient astronomers noticed that Mars moved across the sky with a peculiar motion. This motion was eventually found to be a consequence of a distinctively elliptical orbit that brought Mars to within 206.5 million km (124 million miles) of the Sun (called perihelion) and as far away from the Sun (called aphelion) as 249.1 million km (149 million miles). In contrast, Earth is in a nearly circular orbit about 150 million km (93 million miles) from the Sun.

Because Mars is farther away from the Sun than Earth is, its years are longer: 687 Earth days compared with Earth's year, 365 days.

These differing periods bring Mars and Earth into alignment relative to the Sun (termed opposition) on a 780-day cycle. The differences in type of orbit (elliptical versus nearly circular) mean that the distance between the two planets at opposition is not always the same. The slightly elliptical nature of Earth's orbit (less extreme than that of Mars) plays a role in determining the different distances between the two planets at opposition. Favorable (closer) oppositions occur when Mars is near perihelion and Earth is near aphelion when they line up on the same side of the Sun. Because Earth reaches aphelion in July, the most favorable oppositions occur during the terrestrial Northern Hemisphere summer (Martian Southern Hemisphere summer). For example, at favorable opposition when Mars is at perihelion and Earth is at aphelion, the distance to Earth may be only 55 million km (33 million miles). But if opposition occurs when Mars is at aphelion and Earth is at perihelion, the Mars-Earth distance may be 100 million km (60 million miles). These cycles happen with a period of about 16 years. During a favorable opposition, when Earth and Mars are closest, and under optimum observing conditions, features as small as 150 km (90 miles) can be distinguished from Earth with our largest telescopes.

Mars spins around an axis in ways that are remarkably similar to Earth's spin. The period of this spin is about 24.5 hours, producing a day (called a Sol) that is near the 24-hour Earth day. The tilt of the rotation axis, or obliquity, is about 25°, compared with that of Earth at 23.5°. Like Earth, the tilt of the rotation axis produces seasons on Mars. However, because Mars has a more elliptical orbit, it moves at different rates throughout the year (faster at perihelion than at aphelion), making its seasons considerably different (e.g., currently the winters are 52 days shorter and 30°C [64°F] warmer in the south than winters in the north).

Gravitational effects from the Sun and other planets add a dimension of complication to the annual motion of Mars. Over time, the tug of gravity from these bodies slowly and continuously changes the obliquity, the orbital eccentricity (ellipticity), and the inclination (or tilt) of the orbital plane (Figure 2). Each of these attributes currently changes with its own characteristic period, although there is evidence that these periodic changes may have altered over the history of Mars. The rotation axis of Mars currently precesses with a period of 175,000 years, causing a 180° change in

Table 1

Solar System Planet Data

Parameter	Mercury	Venus	Earth	Mars
Mean distance from the Sun (AU)[a]	0.38	0.72	1	1.52
Sidereal period of orbit (years)	0.24	0.62	1	1.88
Mean orbital velocity (km/sec)	47.9	35.0	29.8	24.1
[miles/second]	[28.7]	[21]	[17.8]	[14.5]
Orbital eccentricity	0.21	0.01	0.02	0.09
Inclination to ecliptic (degrees)	7.00	3.40	0	1.85
Equatorial radius (km)	2,439	6,052	6,378	3,397
[miles]	[1,463]	[3,631]	[3,827]	[2,038]
Polar radius (km)	Same	Same	6,357	3,357
[miles]	[1,463]	[3,631]	[3,814]	[2,014]
Mass of planet (Earth = 1)[b]	0.06	0.82	1	0.11
Mean density (g cm^{-3})	5.44	5.25	5.52	3.94
[pounds/cubic ft]	[337]	[326]	[342]	[244]
Body rotation period (hours)	1,408	5,832[c]	23.93	24.62
Tilt of equator to orbit (degrees)	0	2.12	23.45	23.98

[a]AU indicates one astronomical unit, defined as the mean distance between Earth and the Sun (130 million km [93 million miles]).

[b]Earth's mass is approximately 5.976×10^{26} g (5.976×10^{20} metric tons).

[c]Planet rotation is retrograde (opposite to planet's orbit).

direction of tilt every 87,500 years. The orbital axis precesses with a period of 72,000 years, moving perihelion around to the opposite side of the Sun every 36,000 years. Both effects combine to produce a 25,000-year cycle. This cycle alternately changes the pole that experiences short, hot summers and long, cold winters. Many scientists think that changes in motion cause long-term alterations to the pattern of Martian seasons and can substantially affect its climate.

Motion controls the amount of solar radiation that falls on the surface of Mars at any given place and time, considerably affecting its surface. For example, the orbital eccentricity controls the size of

Jupiter	Saturn	Uranus	Neptune	Pluto
5.20	9.56	19.22	30.11	39.44
11.86	29.46	84.01	164.79	247.68
13.1	9.6	6.8	5.43	4.7
[7.9]	[5.8]	[4.1]	[3.3]	[2.8]
0.05	0.06	0.05	0.01	0.25
1.30	2.49	0.77	1.77	17.17
71,492	60,268	25,559	24,764	1,140
[42,895]	[36,161]	[15,335]	[14,858]	[684]
66,854	54,360	24,973	24,340	Same
[40,112]	[32,616]	[14,984]	[14,604]	[684]
317.8	95.2	14.5	17.5	0.002
1.33	0.69	1.27	1.64	2.0
[82]	[43]	[79]	[102]	[124]
9.92	10.66	17.24	16.11	153.3
3.08	26.73	97.92	28.8	96

temperature differences caused by the differences in distance of Mars from the Sun. The current orbital eccentricity of Mars carries it 20 percent closer to the Sun during perihelion than at aphelion. This results in 45 percent more sunlight and 30°C (64°F) warmer temperature at the subsolar point (point directly beneath the Sun) at perihelion.

Earth has also experienced similar changes in its elements of motion. As they have for Mars, these changes may also have affected the energy balance at Earth's surface, producing periodic climate changes such as the ice ages.

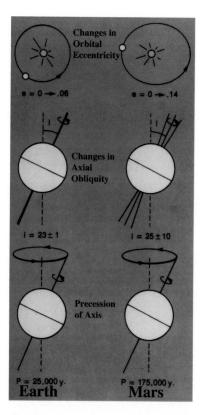

Figure 2. Quasiperiodic changes occur in both the orbital and axial characteristics of Earth and Mars. Variations in these characteristics are much greater for Mars than for Earth. These variations are thought to cause periodic dramatic climate changes on both planets. These periodic changes are brought about by the pull of gravity from other bodies in the solar system. (Modified from J. B. Pollack, Atmosphere of the terrestrial planets, 57–70 *in* J. K. Beatty, B. O'Leary, and A. Chaikin, eds., *The New Solar System,* Cambridge, Massachusetts: Sky Publishing Corporation, 1981)

The Interior

Most of what is known about the interior of Mars is derived from theoretical calculations based on a few observational measurements such as its size, mass, and gravity field; a few types of remote-sensing spectral observations; and the compositions of certain types of meteorites. From these data it is inferred that, like other planets, the interior of Mars differentiated near the time of its formation (4.5 billion years ago) into concentric shells of different chemical composition and physical properties. Mars has a dense iron-rich core, which is surrounded by a lower-density mantle that is overlain by an even lower-density crust.

Initially, the core of Mars may have behaved much like Earth's core, generating a dynamo from convection driven largely by inner-core growth. This dynamo produced a substantial global magnetic field for the first several hundred million years of Martian history. Cooling of the core disrupted this dynamo, eliminating the magnetic field. Currently, Mars has a liquid, conductive outer core and might have a solid inner core like Earth has.

The mantle of Mars is chemically homogeneous, mixed by convection cells that transport deep-seated heat generated during accretion and subsequent core formation. Like Earth's mantle, the Martian mantle has formed concentric shells of different mineralogy caused by the effects of pressure on the components.

The Martian crust formed by melting of the upper mantle. It has been shaped and redistributed by volcanism, impact cratering, fracturing, and erosion. The crust appears to be composed mainly of basalt, like that formed by melting of the upper mantle of Earth. There is also some evidence provided by scientific instruments aboard spacecraft that have visited Mars that other types of rock may be present.

The Surface of Mars

Light and dark markings on the surface have been noted since the early telescopic observations of Mars. Upon closer examination these surface markings are mainly differences in reflectivity caused by surface materials of different composition. The light areas are covered with a thin layer of bright dust, whereas the dark areas are thought to be exposed rock. The gross patterns of the markings have remained nearly constant since their discovery and are believed to be directly related to global wind regimes. However, the details of these markings can change on a short time scale and are related to local and regional storm activity.

In a few cases, regional-scale surface markings seen with ground-based telescopes are related to a prominent topographic feature on the surface (Figure 3). Remarkably, the grandest of all Martian surface features, a global-scale dichotomy in terrain types, is not hinted at by these markings (Figure 4). The Southern Hemisphere of

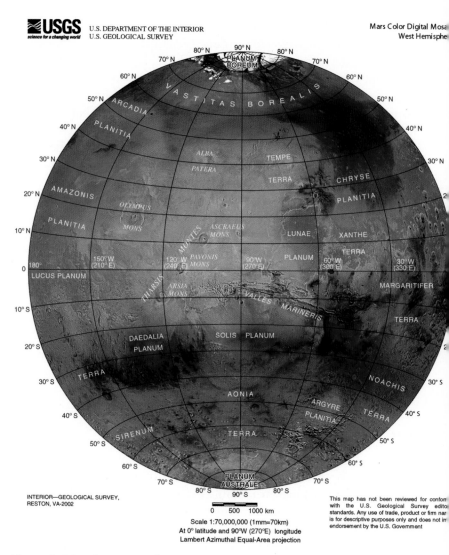

Figure 3. Digital mosaics of Mars projected as hemispheres centered on
90° longitude *(left)* and 270° longitude *(right)*. These mosaics are composed
of hundreds of high-resolution pictures taken by Viking orbiters and other
spacecraft orbiting Mars. (Courtesy U.S. Geological Survey)

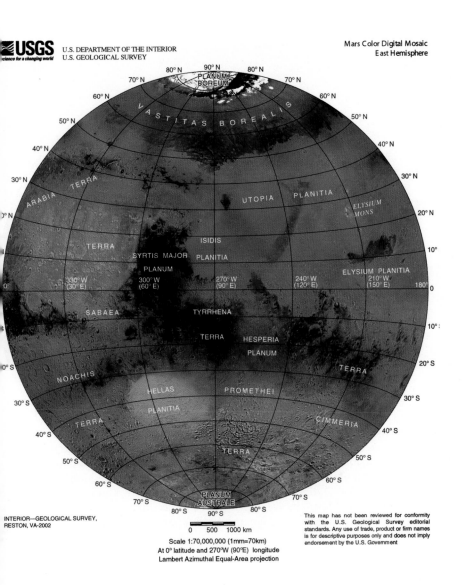

USGS
science for a changing world

U.S. DEPARTMENT OF THE INTERIOR
U.S. GEOLOGICAL SURVEY

Mars Color Digital Mosaic
East Hemisphere

80° N   90° N   80° N
70° N   PLANUM   70° N
BOREUM
60° N   60° N
50° N   V A S T I T A S   B O R E A L I S   50° N
40° N   40° N
30° N   30° N
UTOPIA   PLANITIA
30° N   TERRA   30° N
ARABIA   ELYSIUM
20° N   MONS   20° N
ISIDIS
10°   TERRA   10°
SYRTIS MAJOR   PLANITIA
PLANUM   ELYSIUM PLANITIA
330° W   300° W   270° W   240° W   210° W
(30° E)   (60° E)   (90° E)   (120° E)   (150° E)   180° 0
SABAEA   TYRRHENA
10°   10°
TERRA   HESPERIA
PLANUM
20° S   TERRA   20° S
NOACHIS
30° S   HELLAS   PROMETHEI   30° S
PLANITIA
TERRA   CIMMERIA
40° S   40° S
50° S   TERRA   50° S
60° S   60° S
70° S   PLANUM   70° S
AUSTRALE
80° S   90° S   80° S

INTERIOR—GEOLOGICAL SURVEY,
RESTON, VA-2002

0   500   1000 km

Scale 1:70,000,000 (1mm=70km)
At 0° latitude and 270°W (90°E) longitude
Lambert Azimuthal Equal-Area projection

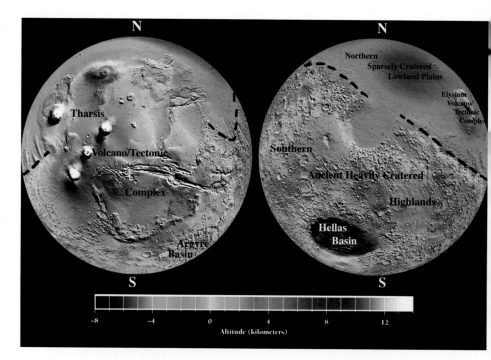

Figure 4. The topography of Mars measured by the Mars orbiter laser altimeter aboard Mars Global Surveyor. The global view on the left is centered on Valles Marineris and shows the enormous volcano/tectonic plateau Tharsis. The view on the right is the opposite side of Mars and clearly shows the dichotomy in terrain types and elevations. The dashed line marks the approximate location of the boundary between the ancient heavily cratered highlands in the south and the sparsely cratered lowlands in the north. Colors indicate elevations, with blue the lowest and red and white the highest. The maps are orthographic projections. (Courtesy NASA and the Mars Global Surveyor Mars orbiter laser altimeter team) (PIA02031)

Mars is composed of ancient, heavily cratered highlands, but the Northern Hemisphere of Mars is covered by sparely cratered low-land plain. In places, a scarp 1 km (0.6 mile) high separates these provinces.

Attesting to the dynamic nature of the Martian interior, two huge volcano-tectonic complexes have developed on Mars. The Tharsis Plateau, only slightly smaller than the crustal dichotomy, is a broad, elongate rise 8,000 km (4,800 miles) long and 7 km (4.2 miles) high that extends northeast-southwest across the equator of Mars. Tharsis is home to huge volcanoes; the most spectacular of these is Olympus Mons, a shield volcano 700 km (420 miles) across and 25 km (15 miles) high. A similar, though smaller, volcano-tectonic complex has also developed in the Elysium region. The Elysium province is enormous by terrestrial standards at about 2,000 km (1,200 miles) across and 5 km (3 miles) high.

Several smaller volcanic centers containing single, large, low-profile volcanoes have developed on fractures associated with the Hellas impact basin. Fields of small volcanoes 2 to 3 km (1.2 to 1.8 miles) across are found scattered around the planet as well.

The crust has been excavated by a number of basin-size impact craters. Two of the largest and most pristine, Hellas and Argyre (1,800 km [1,080 miles] and 800 km [480 miles] diameter, respectively), are located in the southern highlands and are comparable in size with large basins on other planets. Several large impact basins are also buried beneath the north lowland plains, indicating that ancient crust is preserved beneath these plains.

To the east of the Tharsis Plateau is Valles Marineris, an extensive system of interconnecting canyons nearly 5,000 km (3,000 miles) long. In places the main canyon is 700 km (420 miles) across and 7 km (4.2 miles) deep. Its origin is most likely associated with the formation of Tharsis and may be the start of a failed rift zone. Huge outflow channels originate in the lower reaches of these canyons. These outflow channels are thought to be evidence of catastrophic floods.

Numerous valleys dissect the southern highlands, suggesting that the climate of Mars in the past may have been very different than it is today. Some of these valleys are small and form networks, but others are relatively large, solitary features. Much of the water needed to carve these features may have come from subsurface aquifers.

Polar ice caps composed of water ice and carbon dioxide frost sit atop the surface of thinly layered deposits of dust and ice. The layered deposits are several kilometers thick and extend for several hundred kilometers away from the poles. These deposits are thought to have been deposited as a result of cyclic changes in the climate. Likewise, layered deposits found on the floors of craters and in canyons in the equatorial regions are also thought to be the result of climate change events.

The Atmosphere

The Martian atmosphere is thin and cool. It is about 95 percent carbon dioxide, with minor amounts of nitrogen, oxygen, argon, and water vapor. The near-surface atmospheric pressure is generally less than about one-hundredth that of Earth's atmosphere, although there is evidence that periodically it may have been much denser than at present. The temperature of a surface on Mars strongly influences the temperature of the atmosphere. The surface of Mars ranges from 25°C (80°F) in the midlatitudes at midday to −125°C (−193°F) at the South Pole during midwinter. The Martian atmosphere is so thin and transparent that it can hold or transport little heat. There are no oceans on Mars to slow changes in weather brought on by daily and seasonal variations in temperature.

The circulation of the Martian atmosphere is controlled by temperature variations. The strongest surface winds probably occur in lowlands and/or east-facing slopes of large-scale topography. These winds can be strong enough to set particles in motion and raise dust storms that sculpt the surface and leave deposits in many places. Some dust storms can grow to engulf the entire planet.

Like Earth's atmosphere, the Martian atmosphere is layered. The pressure and temperature both decrease continuously up to about 25 km (15 miles), where the temperature remains nearly constant or rises slightly to 110 km (66 miles) altitude. Above that altitude the temperature increases, and low-density gases are lost to space. The loss of components (principally hydrogen) at the top of the atmosphere has had a dramatic effect on changes in the volatile inventory of Mars.

The Satellites of Mars

Two small, dark, potato-shaped satellites, Phobos and Deimos, orbit Mars. These tiny moons are thought to be captured asteroids composed of the same dark materials as carbon-rich meteorites. The orbit of Phobos is unstable and in several hundred million years will cause it to crash into Mars, creating an impact crater over 100 km (600 miles) across.

Chapter 2

Fascination with a Red Planet

Mars was once envisioned as a habitable planet with lush seasonal vegetation and thousands of miles of canals brimming with water. Less than 55 million km (33 million miles) from Earth and a little more than half its size, Mars has always seemed closer and larger to those who have peered at it and followed it through the centuries. Today, after being gazed at and prodded by an armada of spacecraft, Mars is better known for its ruddy-colored boulder-strewn surface, deep waterless canyons, towering volcanoes, and wild dust storms. But it wasn't always so.

Four thousand years ago, soothsayers scanned the heavens for signs to help explain events around them and to predict the future. These early astrologers noticed that every night the stars marched across the sky in fixed patterns, now known as constellations. They divided the sky into twelve zones, each named after the most distinctive constellation in the zone. These were the twelve signs of the zodiac.

But there were five notable exceptions to this orderly parade of stars and constellations. These five celestial bodies looked like stars but behaved differently. They did not belong to any constellation but instead wandered from constellation to constellation. These wandering stars, called planets (from the Greek word *planetes*, meaning wanderer), were even observed to seemingly reverse their direction. Their brightness was observed to change during the course of their wanderings. For the astrologers this erratic behavior aroused considerable attention. Elaborate schemes were developed and claims were made of foretelling the future from the location of these wandering celestial bodies.

Mars had three special traits that drew the attention of ancient observers. Like the other planets, its brightness changed considerably with time and it followed a peculiar path across the sky. But its blood-red appearance most drew the attention of ancient people. It sparked visions of war and carnage. For this reason, the Sumerians christened Mars the god of war more than 3,000 years ago. To the Chaldeans, Mars was Nergal, the master of their battles, their god of death and pestilence. The Egyptians called it Harmakhis or sometimes Har Decher (the red one); to the Persians Mars was Pahlavani Siphir, to the Norse it was Tui (the source of the English word Tuesday), and among the East Indians it was Angaraka (burning coal) or Lohitanga (the red body). The Greeks named it Ares for their god of war, and it was sometimes known as Hercules. But it was the Romans that named it Mars, the name most of us know it by today. In keeping with this traditional association, modern astronomers employ the ancient implements of war, a shield and spear, as the symbol for Mars.

Its color may have given Mars its name, but its strange motion was what gave it a special place in the history of science. As early as 600 B.C., Greek scholars were attempting to explain the odd reversals in direction of Mars as it wandered through the sky. The Greek geometer Eudoxus, in the fourth century B.C., proposed a universe with Earth fixed at its center and the planets and stars fastened to series of concentric spheres around Earth.

Like Eudoxus, astronomer Claudius Ptolemaeus (Ptolemy) from Alexandria, in the second century A.D., also believed Earth to be the center of the universe. To explain the curious movement of the planets, Ptolemy borrowed and expanded on the elaborate geometric system concocted by Hipparchus, a Greek astronomer who lived in the second century B.C. In this system, the planets moved in a complex series of circles, known as deferents, with Earth at a fixed point at the center. To explain the occasional reversal of motion of such planets as Mars, Ptolemy imagined that the planets also moved in small circles, called epicycles. The centers of these epicycles were supposed to travel along the deferents. But to account for the actual position of the planets over time, his system became complicated, requiring more than fifty epicycles.

At odds with this view, the Greek scholar Aristarchus of Samos, in the third century B.C., proposed a revolutionary concept: that the Sun was at the center and the planets revolved around it. But the Earth-centered universe was a strongly entrenched concept, and as a result he was largely ignored.

Until the end of the Middle Ages there was little scientific thought given to Mars. The one exception of note was in the second century B.C. when Aristotle observed that the Moon had passed in front of Mars. This simple observation led him to the conclusion that Mars was "higher up in the heavens" than the Moon and hence farther away. This was a simple, yet important discovery.

Advancement toward understanding the true nature of the universe came to a halt in the Middle Ages. During that time, there was interest in Mars, though only because of its ancient association with bloodshed, war, and carnage. This enhanced its allure and put it at the center of many of the elaborate schemes concocted by astrologists at the time. Progress toward understanding Mars had to wait for the wakening of scientific inquiring that began during the Renaissance.

The Scientific Mars

In 1543, following in the footsteps of Aristarchus, the Polish astronomer Nicolaus Copernicus challenged the ancient Ptolemaic doctrine of Earth's central position. After 30 years of preparation he published his great work *On the Revolution of the Celestial Bodies.* In this work he showed that the motion of the planets could be explained much more simply with Earth rotating about a north-south axis once a day and that Earth and other planets revolved around the Sun. He proposed that each planet orbited in a perfect circle. But to fully account for the observed motions, he was forced to resort to a system of epicycles similar to that of Ptolemy. Copernicus's scheme offered no advantage other than simplicity over the Ptolemy system, so few people supported it.

It was left to a young German astronomer, Johannes Kepler, to make sense of the "ballet of the planets" as Copernicus had called their motion. In 1600 Kepler went to work for Tycho Brahe, the greatest observational astronomer of his time. Though Brahe lived before the telescope was invented, he had developed high-precision astronomical instruments that enabled him to plot the locations of the stars and planets with accuracy far exceeding previous measurements. Because of its peculiar reversals in direction, Mars was the focus of many of his observations. He had produced extensive and accurate plots of the location of Mars spanning nine oppositions, from 1580 to 1600. Kepler used these as a basis for what historians regard as some of the most brilliant scientific work in history. He proved that planets revolve around the Sun and that they do so in elliptical orbits, instead of perfect circles. This discovery had far-reaching effects, laying the foundation for Kepler's three laws of motion, as well as setting the stage for Newton to develop his theory of universal gravity, all of which are essential for navigating spacecraft.

In the same year that Kepler made his discovery, 1609, Galileo Galilei pulled the Red Planet a bit closer. He pointed the first telescope into the sky and brought Mars into sharp focus, recording his observations in his book, *The Sidereal Messenger.* Mars was no longer just a flickering red light in the night sky. He peered at Mars, trying to observe its phases predicted by the work of Kepler and Copernicus. But because his telescope was still quite primitive, Galileo failed. Writing a friend, he proclaimed, "I dare not affirm that I am able to observe the phases of Mars; nonetheless, if I am not mistaken, I believe I have seen that it [Mars] is not perfectly round."

The Italian astronomer Francisco Fontana fared only slightly better in his observations of Mars. In 1636 he produced the earliest known drawing of Mars, and in 1638 he was able to observe the phases of Mars (Figure 5), noting that "the disk of Mars is not uniform in color." But his telescope was nearly as primitive as Galileo's. As a result his drawings showed nothing that resembles the actual markings on Mars and hence they are of only historical significance.

As telescopes were refined, astronomers began noting more and more similarities between Earth and Mars. In 1659 Dutch mathematician and physicist Christian Huygens found that Mars, like Earth, rotated about a north-south axis and that the length of its day was very close to Earth's 24-hour period. (Today the exact length of one Martian day is 24 hours, 37 minutes, and 22 seconds.) Huygens's telescopic observations inspired him to draw maps of Mars that included what is known today as the dark triangle of Syrtis Major (Figure 6).

Italian astronomer Giovanni Cassini took another important step, reporting the sighting of white polar caps on Mars. Forty-seven years later his nephew, Giacomo Maraldi, observed that not only did the polar caps discovered by his uncle change size but also that the center of each cap did not correspond to the Martian poles. The same holds true on Earth, where the Arctic and Antarctic ice caps are not centered on the north and south rotational poles.

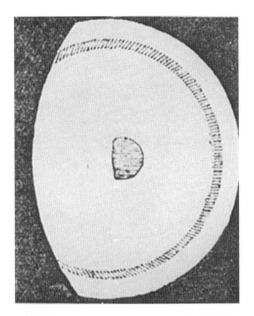

Figure 5. Drawing of Mars made by Fontana in 1638. The drawing shows the gibbous phase of Mars.

Following Maraldi, in the late 1700s, the musician turned astronomer Sir William Herschel noted more similarities of Mars to Earth. He became convinced that their surface conditions were so similar that both were teeming with populations of living creatures. One of these similarities was seasons on Mars. As part of his observations, Herschel carefully measured the orientation of the poles of Mars and discovered that the rotation axis is tilted at about 25°, nearly the same angle as Earth's tilt (23.5°). He reasoned that as a consequence of this inclination Mars had four seasons, much like Earth, but twice as long as Earth's because a Martian year is equal to 687 Earth days. This, combined with Maraldi's observation that the bright polar caps grew and shrank with the season, led Herschel also to conclude that these caps consist of thin sheets of snow and water ice.

Herschel had also noted light and dark markings on the surface of Mars and was convinced that most of the dark areas were bodies of water, just as the dark areas on the Moon were thought to be oceans or seas. The exception was the dark band that appeared near the edge of the polar caps in spring and extended toward the equa-

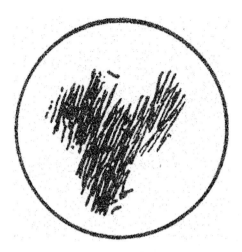

Figure 6. Drawing of Mars by Huygens in 1659 showing Syrtis Major. South is at the top, a convention adopted by users of telescopes (which invert the view).

tor as summer came on. He believed this to be a product of melting of the ice and snow of the caps. But in contrast to the behavior of the dark markings, changes in the bright markings, especially in the high latitudes, were attributed to changes in the atmosphere. He thought these markings were "clouds and vapors floating in the atmosphere of the planet." It was found later that these changes were most likely due to seasonal changes of dust cover on the surface.

The Geographic Period

In the early part of the 1800s, new, larger telescopes offered substantial improvements in the ability of astronomers to see details on Mars. As a result, the 1830 opposition saw a resurgence of interest in Mars. This started what the French astronomer Camille Flammarion called the "Geographic Period" of investigations of Mars. In some cases, mapping of the surface was pushed beyond the limits of resolution of the telescopes into the realm of optical illusions and imagination. Some of these observations play heavily in the folklore of Mars, fostering bizarre interpretations for the cause of these features.

At the start of this period, no complete map of the planet existed until German banker William Beer, working with the astronomer Jo-

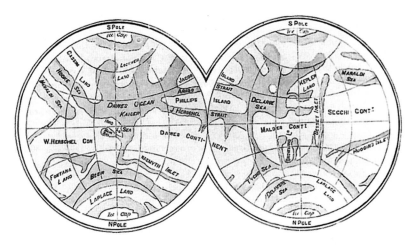

Figure 7. Proctor's map of Mars with named features, published in 1871. Later astronomers discarded the names on this map. The "Kaiser Sea" centered in the left hemisphere now bears the name Syrtis Major.

hann H. von Mädler, painstakingly assembled one in 1840. They initially produced a sketch map that contained few features that are recognizable as real surface markings on Mars. However, most notably this map defined the latitude and longitude system (areographic) on Mars. The Beer-Mädler system introduced in 1840 is essentially the same as that currently used by astronomers and mapmakers.

Mappers also recognized the need to assign names to features they mapped, making those features easier to identify and discuss. Early sketch maps contained names, but it was the English astronomer Richard Proctor in 1867 who made the first systematic attempt at a Mars nomenclature (Figure 7).

Flammarion was critical of Proctor's maps and nomenclature. He sketched his own map and introduced his own nomenclature. In doing so, he remarked that Proctor's nomenclature had not found wide acceptance because it had "given too much recognition to the astronomers of his own country, and for having repeated the same names." For example, Proctor had used the name Dawes, his fellow countryman, for six features—a serious source of confusion. But Flammarion's names, like those assigned by Proctor, are also of only historical significance—they became the victims of detailed observations from greatly improved telescopes during the opposition of 1877.

It should be noted that a second "Geographic Period" in Mars investigations began in the 1960s. It was led by Raymond Batson, Sherman Wu, and David Scott of the U.S. Geological Survey; Merton Davies of the Rand Corporation; and David Smith of the NASA Goddard Space Flight Center. During that period, Mars was mapped in great detail using pictures and other measurements from spacecraft that visited it.

Discovery of the Moons of Mars

The summer of 1877 was a particularly good time for observing Mars. Mars was the closest it ever gets to Earth during the 1877 opposition, and at the same time telescope making had become a fine art that produced enormously powerful instruments.

One of the most important discoveries during this favorable opposition was the discovery by Asaph Hall of the U.S. Naval Observatory in Washington, D.C., of the two tiny satellites of Mars, Phobos and Deimos. Hall wrote of his discovery in August 1877 that, "on the night of the 11th, and at half past two o'clock, I found a faint object . . . a little north of the planet, which afterward proved to be the outer satellite." Following up on that initial observation he found "on August 16, the object was again found . . . and the observation . . . showed that it was moving with the planet . . . . On August 17, while waiting and watching for the outer satellite, I discovered the inner one. The observations on the 17th and 18th put beyond doubt the character of the objects."

More will be said later about the discovery of these satellites.

The Canals of Mars

Looking through a new high-powered telescope and taking advantage of the choice viewing in 1877, Giovanni Schiaparelli, director of the Brera Observatory in Milan, Italy, plotted the locations of sixty-five bright and dark features on Mars. He gave them historical and mythical Latin names, most of which are still in use today. The bright surface areas he called Hellas and Amazonis for arid, wind-blown, terrestrial deserts and other bright areas after Earth's own Arabian and Syrian deserts. The dark regions were called Boreum Mare and Syrtis Major after Earth's North Sea and the Mediterranean Sea.

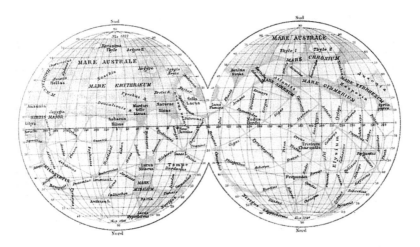

Figure 8. Mercator projection map of Mars made by Schiaparelli.

Of particular note, Schiaparelli's hand-drawn maps depicted dark intersecting lines, which he believed to be strange networks stretching hundreds of miles across Mars (Figure 8). He borrowed the Italian astronomer Father Secchi's term *canali* to refer to the lines in these networks. With this casual use of the term *canali*, Schiaparelli inadvertently laid the foundation for the birth of an elaborate modern mythology about Mars. In Italian *canali* means "channels"; however, it was later mistranslated to mean "canals." And how did those canals get there? Someone had to build them, of course.

In an article in the French astronomical journal *L'Astronomie* in 1882, Schiaparelli wrote:

There are on this planet, traversing the continents, long dark lines which may be designated as canali, although we do not know what they are. Those lines run from one to another of the somber spots that are regarded as seas, and form, over the lighter or continental regions, a well-defined network. Their arrangement appears to be invariable and permanent; at least as far as I can judge from four-and-half-years of observation. In 1879 a great number were seen which were not visible in 1877, and in 1882 all those that had been seen at former oppositions were found again, together with new ones. Sometimes these canals present themselves in the form of shadowy and vague lines, while on other occasions they are clear and precise, like a trace drawn with a pen. I am absolutely certain of what I have observed.

The discovery of these linear features on Mars is frequently attributed to Schiaparelli. But in a report to the Society of Italian Spectroscopists in 1878 he pointed out that he was far from the first to see these features. He wrote, "Some people have been inclined to doubt the existence of the canali since they have not seen them. Yet many of the canali are not new and have already been seen by such excellent astronomers as Kaiser, Lockyer, Secchi, Green, etc." He could have also added Dawes and Proctor, as well as the German astronomer Johannes Schroeter, to the list. Though the others may have seen them first, it was Schiaparelli who called attention to these features. He gave them specific names and recorded over a hundred of them.

It was almost a decade before sightings of these canals became common by other astronomers. Astronomers began to see more and more details as they observed these features. For example, in 1892 William Pickering from Harvard Observatory spotted a canal running across Mare Erythraeum, a dark region on Mars that Schiaparelli thought contained an ocean. Recognizing that a canal cannot run across an ocean, Pickering reasoned that the dark areas were not bodies of water at all. He declared them to be vast regions of vegetation.

No one was more passionate about Mars than Percival Lowell. He was so captivated by the Red Planet that he devoted his life to studying it. Lowell was a member of a distinguished and wealthy family in Boston and an explorer of and writer about the Far East. Because of his driving interest in planetary astronomy, he founded the Lowell Observatory near Flagstaff, Arizona, to take advantage of the high altitude and clear air. Within a short time after founding the observatory in 1894, he began to see Schiaparelli's canals, eventually identifying more than five hundred of them, more than four times the number discovered by Schiaparelli.

Lowell realized that the canals were too wide to be the waterways originally described by Schiaparelli. He reasoned that the networks must actually be agricultural regions, covered by vegetation and irrigated by invisible small canals that carried seasonal meltwater flowing from the polar caps. Lowell was sure that there was no natural explanation for the canals and that they must have been constructed by "intelligent creatures, alike to us in spirit, though not in form." It was this firm opinion that the canals were the work of intelligent beings and his ability to popularize his ideas that sparked an enormous tide of public interest in Mars at the end of the nineteenth century. During that time, Lowell wrote three books about Mars: *Mars* in 1895, *Mars and Its Canals* in 1906, and *Mars the Abode of Life* in 1908, each more enthusiastic about the idea of life on Mars than the previous one.

On theories of how the canals formed on Mars, it appears that Lowell had captured the imagination of more than the general population. In 1893, before Lowell started his observations, the man who made canals a popular idea, Schiaparelli, wrote, "It is not necessary to suppose them [*canali*] to be the work of intelligent beings, and . . . we are now inclined to believe them to be produced by the evolution of the planet." But in 1897, as the fervor grew about the canals, and apparently under Lowell's influence, Schiaparelli wrote, "[The] arrangement [of the *canali*] present an indescribable simplicity and symmetry, which cannot be the work of chance." The idea of life on Mars is, indeed, a seductive concept.

With Lowell stoking the fires of belief that Mars harbored civilization, it is easy to understand why science fiction and adventure writers such as H. G. Wells, Arthur C. Clark, Ray Bradbury, and many others used Mars as a backdrop for their fantastic tales. Inspired by Lowell, these writers produced a delightful series of fictional accounts of the inhabitants of Mars and their exploits—scaring, and inspiring, generations of Earthlings. These wonderful writings are outside the scope of this book but are well worth exploring.

Peering through a telescope twice as powerful as Schiaparelli's, Edward E. Barnard of the Lick Observatory saw none of the lines that Lowell and Schiaparelli claimed were waterways. In a letter to his colleague Simon Newcomb of the U.S. Naval Observatory, Bernard wrote, "I have been watching and drawing the surface of Mars. It is wonderfully full of detail. There is certainly no question about there being mountains and large, greatly elevated plateaus. But to save my soul, I can't believe in the canals as Schiaparelli draws them. I see details where he draws none. I see details where some of his canals are, but they are not straight lines at all. I verily believe all the verifications that the canals depicted by Schiaparelli are a fallacy. . . ."

These were serious charges by Barnard, and he was severely criticized by Lowell and his colleagues. Lowell's response was direct and confident: "In our exposition of what we have gleaned about Mars, we have been careful to indulge in no speculation. Our conclusion is this: that we have in these strange features, while the telescope reveals to us, witness that life, and life of no mean order, at present inhabits that planet."

Most astronomers in the early part of the twentieth century doubted Lowell's conclusions. They suspected the network of lines that he so confidently claimed were canals were in reality optical illusions and a product of his imagination. After Lowell's death, in 1916, the romance of Lowell's Mars rapidly gave way to the "real" Mars. Astronomers' observations pointed toward a planet not only uninhabited by intelligent beings, but one that they generally regarded as uninhabitable.

Evidence for the existence of canals began to lose credibility rapidly as astronomers turned their new, large, powerful telescopes toward Mars to search for signs of Lowell's canals and found none. When George E. Hale and his associates pointed the 1.5-m (60-inch) telescope at Mount Wilson toward Mars in 1909, they reported "not a trace of geometric structure on the planet, nor any narrow straight canals." That year, the French astronomer E. M. Antoniadi also found that "The planet revealed a prodigious and bewildering amount of sharp or diffused natural, irregular detail, all steadily; and it was at once obvious that the geometrical network of single and double canals discovered by Schiaparelli was a gross illusion." It was becoming clear to most astronomers that Lowell's canals were not real.

In 1913 the British astronomer Edward W. Maunder dealt Lowell's canals another blow. In a simple experiment, he showed how easy it was for the human eye to connect scattered spots into a line. He took a large piece of paper and drew a random assortment of lines and circles. Then he asked students to draw what they saw. Comparing the drawings of students who sat closest to the paper with those who sat farthest away showed that the drawings made from farthest away contained imaginary lines that connected individual circles and lines. But there were still a few diehards who persistently claimed to see canals up to the time the issue was finally settled by the early Mariner spacecraft.

To the end, Lowell was a staunch believer in a civilization of Martians who had constructed an intricate network of irrigation canals to feed their thirsty crops throughout the vast deserts of Mars. Because of this belief and his talents as a writer, Lowell left a legacy—

Mars mania. During his lifetime, he had single-handedly pushed Mars to the forefront of public popularity and scientific interest.

New Mars

Early in the twentieth century, at the same time visual observations were making important breakthroughs in resolution, there was an explosion in the development of new astronomical instruments and techniques. These allowed astronomers to do much more than sketch Mars or take its picture. These new techniques could measure what conditions were really like on Mars: its temperature, atmospheric pressure, and composition of the surface and atmosphere. For scientists, this was the beginning of an exciting era of scientific investigation of Mars when conditions on the surface were measured instead of interpreted from visual observations.

Immediately after the invention of the method of spectral analysis, in which analysis of how the atmosphere/surface of Mars absorbs certain wavelengths of sunlight provides information about the composition of the atmosphere/surface, astronomers recognized its potential for probing Mars. In 1862 the English astronomer William Higgins began the first study of the spectra of Mars in an effort to measure its atmospheric pressure. He managed to detect reflected sunlight, only proving that Mars does not glow. Undaunted, astronomers doggedly continued to develop spectral analysis techniques (and other promising new remote sensing methods) and by the 1920s had refined these methods to a point where they started to produce reliable results.

Intent on proving that Mars was, indeed, an abode of life, in 1908 Lowell turned these new tools toward Mars in an effort to measure the pressure of the atmosphere. He measured its pressure as about 87 percent of that of Earth. We know now that this is ten times greater than the actual value. The method he used, based on scattering of light by the gases in the atmosphere, would have given the right answer, but he underestimated other important contributors to the scattering of light, such as the amount of dust in the atmosphere. Even though he got the wrong answer, his attempt is notable and ushered in a new era in understanding Mars.

Like Lowell, other astronomers continued to struggle for an accurate measurement of Martian atmospheric pressure, but none was particularly successful for the next half century. Most used spectroscopy techniques that measured the pressure of only one component or element in the atmosphere. As a result, without complete knowledge of the composition of the atmosphere it is impossible to calculate the total pressure. This type of compositional information was not available until spacecraft visited Mars in 1965. In 1963 David Kaplan and his colleagues at the Jet Propulsion Laboratory came very close to determining the correct value of approximately 5 mbar (0.07 pounds/square inch). They collected a detailed spectrum of Mars, found water vapor, and calculated the pressure of carbon dioxide to be 4 mbar (0.06 pounds/square inch), very close to the actual value. But because they were unaware that the Martian atmosphere is composed mainly of carbon dioxide and that other components, such as nitrogen, are only found in small amounts, they concluded that the total pressure was about 25 mbar (0.37 pounds/square inch). However, the prize for the first truly accurate measurement of the atmosphere from Earth goes to Louise Young, an American astronomer. In 1971 she used the newly invented Fourier spectrograph to measure the major components and determined a value of 5.16 mbar (0.075 pounds/square inch). Although this was after spacecraft had visited Mars and already provided an accurate measurement, it proved that the Earth-based techniques used by early astronomers were accurate.

Spectral analysis also began to reveal what the surface of Mars was really like. Thermal radiation, heat energy absorbed from sunlight by the surface and atmosphere and reradiated back to space, if measured properly, can be used to calculate the surface temperature. Such measurements were made at Lowell Observatory during the 1920s. The temperatures derived from these measurements indicated that the surface of Mars is a cold place, averaging $-40°C$ ($-40°F$), as compared with the average temperature of Earth of $15°C$ ($59°F$). The poles were even colder, at $-70°C$ ($-94°F$), and the warmest place, directly under the Sun, was about $10°C$ ($50°F$). Later, in 1954, W. M. Sinton and J. Strong, also at Lowell Observatory, using a much larger telescope and newer instruments, updated these measurements, finding that the warmest point on the surface was actually $25°C$ ($77°F$).

These temperatures had implications for the composition of the polar ice. They seemed to support the assumption of Herschel and Cassini that the Martian ice caps were made of water ice. These measurements showed that the temperature at the poles is always above $-100°C$ ($-148°F$) and therefore is consistent with water ice and is too warm for carbon dioxide ice to exist ($-125°C$ [$-193°F$]). This composition was also consistent with the spectral measurements by the Russian astronomer Vassili Moroz and the Dutch-American astronomer Gerard Kuiper. There were still a few die-hard astronomers, such as the Americans J. Joly, G. J. Stony, and A. C. Runyard, who questioned if the caps were composed of water ice. They reasoned that carbon dioxide must be a major component of the ice because of lack of water in the atmosphere, although they could not explain how carbon dioxide ice could exist at such a warm temperature.

In 1965 Mariner 4 provided the key to this dilemma when it measured the atmospheric pressure. This gave Robert B. Leighton and Bruce C. Murray of the California Institute of Technology and Conway Leovy of the National Center for Atmospheric Research an accurate value of pressure to use in theoretical calculations of the actual polar surface temperature. Their calculations indicated that the low pressure and composition of the carbon dioxide–rich atmosphere combined to depress the surface temperature below the frost point of carbon dioxide at the poles. It was then left to Murray's student, Hugh Kieffer, to explain why the spectral measurements of Moroz and Kuiper showed the caps to be made of water ice, instead of carbon dioxide ice. Kieffer found in his laboratory measurements that even small amounts of water masked the spectrum of carbon dioxide in mixtures of water and carbon dioxide ice. Consequently, because water ice is expected to condense at the poles along with carbon dioxide ice, though present the carbon dioxide ice would not be detected.

Astronomers have long recognized that, besides the canals, most of the Martian surface is a mottled collection of bright and dark markings. Many early observers noted seasonal changes in the dark markings and described their color as being greenish or bluish. The most notable of these changes was the "Wave of Darkening." This darkening of the surface allegedly progressed outward slowly from the springtime edge of the polar caps and reached across the equator. Some claimed the darkening was due to growth of the vegetation cover in those areas. In 1938, using spectroscopy, the Canadian astronomer Peter Millman gave this interpretation a serious blow. He compared the spectrum of light reflected from the dark areas on Mars with that for leafy green vegetation on Earth and found them to be very different. He considered this clear evidence that the dark marking could not consist of vegetation.

Some ideas die hard, and that of abundant life on Mars was one of these. A few astronomers, mostly at Lowell Observatory, continued to search for signs of life on Mars. One of these astronomers was W. M. Sinton. In 1954 he claimed he had collected infrared spectra of the dark areas that showed evidence of organic compounds on Mars. He declared that this was evidence that the wave of darkening was, in fact, caused by the seasonal growth of vegetation. He quickly issued a report called "Spectroscopic Evidence for Vegetation on Mars," but later withdrew it. After further studies with his colleague Donald Rea of the University of California, he and Rea found that the original spectra he had collected were contaminated by the spectra of molecules of heavy water in Earth's atmosphere. By coincidence, the spectral signature of heavy water (water containing a deuterium or heavy hydrogen atom) found in Earth's atmosphere is similar to the spectral signature of certain organic molecules that are diagnostic of life. To their credit, Sinton and Rea later confirmed that there appeared to be no spectroscopic evidence for complex organic materials on Mars.

Following in Millman's footsteps, the French astronomer Audouin Dollfus and the American astronomers Thomas McCord at the Massachusetts Institute of Technology and Peter Boyce (no relation to the author) of Lowell Observatory also analyzed spectra of these dark areas and came to another surprising conclusion. They found that the dark areas were not greenish or bluish but merely less red than the bright areas. Their observations indicated that the "Wave of Darkening" is mainly caused by brightening of the bright areas on the surface rather than darkening of the dark areas. Consequently, the "Wave of Darkening," like the canals, is an illusion and is really a wave of brightening caused by seasonal changes in the thickness of bright-colored dust that covers the surface of Mars.

It was clear by the 1960s that the surface of Mars is a cold, dry, harsh place with little possibilities for life to thrive. In 1961 the National Academy of Sciences convened a panel of experts to advise NASA on the presumed evidence for water ice at the Martian poles and water vapor circling the planet. Although Sinton's evidence for vegetation was not accepted, the report concluded: "The evidence taken as a whole is suggestive of life on Mars. In particular, the response to the availability of water vapor is just what is to be expected on a planet which is now relatively arid, but which once probably had much more surface water. The limited evidence we have is directly relevant only to the presence of microorganisms; there is no valid data for or against the existence of larger organisms and mobile animals."

Shortly thereafter, biologist Norman H. Horowitz summed up the attitude of most workers in this field: "if we admit the possibility that Mars once had a more favorable climate . . . we cannot exclude the possibility that Martian life succeeded in adapting itself to the changing conditions and remains still there." Willard Libby, the Nobel Prize winner for work leading to his development of radiocarbon dating, added to this notion. He commented, "Intelligent life as we know it can hardly be expected on the surface of . . . Mars . . . for two reasons—the extreme swings in temperature and the killing nature of unfiltered sunlight. So the possibility of life . . . would appear to be restricted to subterranean forms requiring no atmosphere and no sunlight. Our studies on meteorite matter show that considerable quantities of organic matter, presumably primeval, exist, and we can think therefore of anaerobic bacteria living off of this matter. . . . Anaerobic subterranean life may be chemically possible on . . . Mars."

Over the past four thousand years our perception of Mars has changed many times. It began in myth, with a red dot in the night sky that was thought to control human destiny. As the newly invented telescope drew it closer, it became an Earth-like abode teeming with masses of intelligent creatures that tended their desert farms along great systems of canals. As the twentieth century began, the Earth-like environment and the great population of Martians had begun to give way to a harsh, cold, barren world devoid of any kind of life. But our impression of Mars was about to change again as a result of the exploration of Mars by spacecraft.

Chapter 3

The Difficult Target

Events during the summer of 1965 changed our view of the face of Mars forever. On 14 July, Mariner 4 became the first spacecraft to explore Mars successfully. There were unsuccessful attempts before it: five Soviet craft had failed, most even to leave Earth orbit, and a stuck shroud on Mariner 3 prevented its flyby of Mars (Table 2).

The 261-kg (575-pound), windmill-shaped Mariner 4 spacecraft (Figure 9) began its journey from Cape Canaveral, Florida, to Mars on 28 November 1964 at 9:22 A.M. Eastern Standard Time. It was neatly packaged inside a newly designed metal shroud to reduce aerodynamic resistance during launch and to protect it from heat damage as it passed through Earth's atmosphere. Shortly after blast-off on an Atlas-Agena rocket booster, the shroud dropped away and the solar panels unfolded and went to work. All systems were go for a 187.6-million-km (113-million-mile) journey through space. Mariner 4 was off to a good start.

Mariner 4 carried a host of scientific instruments to study Mars and the space environment between Earth and Mars. For example, a magnetometer was mounted on the spacecraft to measure the magnetic field characteristics between Earth and Mars and to detect the existence of a Martian magnetic field. The experiments that had the greatest impact were the camera (it exposed Mars as it had never been seen before) and the radio-occultation experiment (it provided the first direct measurement of the Martian atmosphere when Mariner 4 swung around the planet and its radio signal passed through the atmosphere).

Flying to Mars was not a straight shot. On 4 December, shortly after launch, Mariner 4's trajectory was so far off course that if it had not been corrected, it would have missed Mars by 256,000 km

(154,000 miles). As terrestrial mariners had done throughout history, Mariner 4 used a guide star, Canopus, as a point of reference to navigate and orient the craft in space. Using this star as a reference, Mission Control rolled and turned the spacecraft in search of the guiding light from Canopus. After nearly a full day of maneuvers, nitrogen jets were fired, putting Mariner on a course that would take it to within 9,846 km (5,208 miles) of the surface of Mars.

On 14 July 1965, after 228 days in flight, Mariner 4 reached the orbit of Mars. This rendezvous with Mars had been no easy feat. One Mariner scientist compared the accuracy that brought the spacecraft this close to Mars with rolling a strike in a bowling alley reaching from Los Angeles to San Francisco.

The actual route to Mars had been calculated half a century before by Walter Hohmann, a German mathematician. The most economical way to send a probe to Mars, Hohmann figured, was to follow a long, curved orbit that would intersect with a point in the orbit of Mars at the exact time that Mars itself arrived at that point (Figure 10). For Mariner 4 that meant a 228-day trip covering a distance of 455 million km (273 million miles). All that way, to arrive at Mars at a straight-line distance of 187.6 million km (113 million miles) from Earth!

About 78 minutes after Mariner 4's closest approach to the planet, its trajectory took it behind Mars, where it remained in darkness for 45 minutes. During that time all radio transmission signals to Earth stopped. As soon as it reached the sunlit side, we got our first close-up look at the surface of another planet (Figure 11), and Mars instantaneously became more than a twinkle in the evening sky. Mariner 4 continued on in its trajectory past Mars and into a huge orbit around the Sun.

A Moon-Like Planet

As the first photograph slowly filled the television monitors in Mission Control, scientists were both stunned and disappointed. It was not the shot they had been waiting for. The dark hazy image had captured a 280-km (168-mile) heavily cratered stretch of an area that was called Memnonia on astronomer's maps. Where were the canals? Where was the water? The 1 percent of the surface photographed by Mariner in this sequence of twenty-two pictures was covered with mountains, valleys, and craters like those on the Moon.

Table 2

Mars Missions

Mission	Launch Date	Fate
Mars 1960A (USSR)	10 October 1960	Failed to reach Earth orbit
Mars 1960B (USSR)	14 October 1960	Failed to reach Earth orbit
Mars 1962A (USSR)	24 October 1962	Failed to leave Earth orbit
Mars 1 (USSR)	1 November 1962	Communications failed
Mars 1962C (USSR)	4 November 1962	Failed to leave Earth orbit
Mariner 3 (USA)	5 November 1964	Shroud stuck
Mariner 4 (USA)	28 November 1964	14 July 1965 flyby at 9,920 km (5,952 miles)
Zond 2 (USSR)	30 November 1964	Communications failed
Mars 1969A (USSR)	27 March 1969	Failed to reach Earth orbit
Mariner 6 (USA)	24 February 1969	1 July 1969 flyby at 3,437 km (2,062 miles)
Mariner 7 (USA)	27 March 1969	5 August 1969 flyby at 3,551 km (2,134 miles)
Mariner 8 (USA)	8 May 1971	Failed to reach Earth orbit
Kosmos 419 (USSR)	10 May 1971	Failed to leave Earth orbit
Mars 2 (USSR)	19 May 1971	Lander crashed
Mars 3 (USSR)	28 May 1971	Lander transmitted 20 seconds
Mariner 9 (USA)	30 May 1971	14 November 1971 in Mars orbit
Mars 4 (USSR)	21 July 1973	Failed to enter Mars orbit
Mars 5 (USSR)	25 July 1973	12 February 1974 in Mars orbit
Mars 6 (USSR)	5 August 1973	Lander transmitted descent data only
Mars 7 (USSR)	9 August 1973	Lander missed planet
Viking 1 (USA)	20 August 1975	19 June 1976 in Mars orbit/20 July 1976 landed on Mars
Viking 2 (USA)	9 September 1975	7 August 1976 in Mars orbit/3 September 1976 landed on Mars
Phobos 1 (USSR)	7 July 1988	Failed to reach Mars
Phobos 2 (USSR)	12 July 1988	29 January 1989 acquired data; 27 March 1989 lost contact
Mars Observer (USA)	25 September 1992	Failed on Mars approach
Mars Global Surveyor (USA)	7 November 1996	11 September 1997 in Mars orbit

Mission	Launch Date	Fate
Mars '96 (Russia)	16 November 1996	Failed to reach Earth orbit
Mars Pathfinder (USA)	4 December 1996	4 July 1997 landed on Mars
Nozomi (Japan)	3 July 1998	Initial problems, January 2004
Mars Climate Orbiter (USA)	10 December 1998	Failed to enter Mars orbit
Mars Polar Lander (USA)	3 January 1999	Lander crashed
Deep Space 2 probes (USA)	3 January 1999	Probes crashed
Mars Odyssey (USA)	7 April 2001	23 October 2001 in Mars orbit

In ten of the images seventy craters were counted that measured 4.8 to 20 km (3 to 12 miles) in diameter.

All records for long-distance communication in space were broken by Mariner 4. Its weak radio pulses were transmitted to Earth over a 187.6-million-km (113-million-mile) span at a signal strength of less than one-tenth of a billionth of a billionth of a watt. Each of the twenty-two images was made up of two hundred lines and took 8½ hours to transmit to receiving stations on Earth. The signals were received line by line at a rate of one every 2½ minutes.

The signals were electronic bits of information that translated into 40,000 binary digits representing sixty-four possible shades of black, white, and gray; zero was the brightest, sixty-four the darkest. Once the numbers were decoded by computers, they were transferred to magnetic tapes for insertion into a device called a film recorder. The recorder scanned the film with a beam of light, which changed the number to its corresponding shade.

The images were a disappointment to most everyone: there were no canals. A statement from NASA accompanied each photograph released to the public; it stated that no evidence of "canals" or sea basins on Mars were visible, even though maps and charts contained many of these controversial features.

"According to the charts, one would expect to see so-called canals crossing each of these frames. I have not been able to see any evidence so far of narrow, uniform linear features that could be interpreted as canals," said Robert B. Leighton, physicist-astronomer and chief experimenter on the Mariner 4 photo experiment. What

Figure 9. The Mariner 4 spacecraft was a windmill-shaped spacecraft 7 m (22 feet) across and 3 m (9½ feet) high. The total weight was 261 kg (575 pounds). (Courtesy NASA)

appeared instead of canals were a number of features that resembled small craters, odd-shaped depressions, and ridges.

In their reports to President Lyndon Johnson on the details of Mariner's findings, scientists noted that Mars was more like the Moon than like Earth. Its surface was ancient and could be two to five billion years old. They added that Mars was a hostile planet with an environment similar to that found 7 km (4.2 miles) above Earth. Clearly, Mars did not appear to be a promising place to live.

The idea of life existing on Mars was further challenged when the radio-occultation experiment on board Mariner 4 found the atmosphere to be less than 1 percent of the pressure of Earth's atmosphere and largely carbon dioxide. Because the diameter of Mars is a little more than half of Earth's and it has about one-tenth the mass, Mars lacks the essential gravity to retain much of an atmosphere.

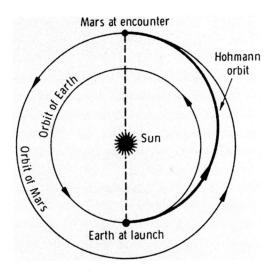

Figure 10. An ideal Hohmann transfer trajectory to Mars. This is the trajectory used by nearly all missions to Mars. (Courtesy NASA, from S. Glasstone, *The Book of Mars*, NASA Special Publication 179, 1968)

Figure 11. One of the highest-resolution pictures of Mars taken by Mariner 4 on 15 July 1965. This picture is of the region of Atlantis and has a resolution of about 3 km (1.8 miles). It clearly shows that the surface of Mars contained numerous impact craters, much like Earth's Moon. (Courtesy NASA) (PIA02980)

Mariner 4 did not detect a magnetic field around Mars. The lack of a Martian magnetic field meant that there was no protective layer between the surface of the planet and bombarding cosmic rays: an unlikely environment for sustaining life. Earth's magnetic field provides partial protection for life by deflecting some of the incoming cosmic rays.

After Mariner 4's visit, Mars still remained the most tantalizing object in the solar system to astronomers and planetary scientists. Although the twenty-two-picture sequence did not capture images of a thriving civilization, it shed light on what was then thought to be the real Mars: a barren desert world that had changed little for eons.

Secrets Slowly Revealed

Beginning in 1960, the Soviets had made five attempts to reach Mars: Project 1M in 1960 (two spacecraft), Mars 1 in 1962, and Project M-69 (two spacecraft) in 1969. All had failed. The Americans continued to have better luck with Mariners 6 and 7 in 1969: two more flyby spacecraft. Nearly identical to Mariner 4, these Mariners were only slightly larger and carried more instruments.

Mariner 6 was launched on 24 February 1969 and Mariner 7 one month later. Mariner 6 flew over the equator to study the surface and atmosphere while its twin observed the Southern Hemisphere. Between the two of them they took 202 photographs before leaving Mars, mapping 9 percent of the planet. Some differences from the Mariner 4 views did appear: large expanses of chaotic terrain and jumbled blocks of crust. The cameras picked up clouds over the south polar cap, and infrared instruments recorded temperatures as low as $-125°C$ ($-193°F$). This result implied that the polar cap was composed mainly of frozen carbon dioxide and not frozen water. Martian snow was definitely different than our own.

Mariners 8 and 9 were to be an improvement on their predecessors; instead of flying by Mars, they would go into orbit about Mars and stay awhile. Mariner 8 met with disaster on 8 May 1971 when shortly after takeoff its launch vehicle failed and it plunged into the Atlantic. Mariner 9 (Figure 12) would have to be reprogrammed to assume the tasks of both spacecraft.

Mariner 9 carried with it more powerful and higher-resolution instruments than the previous Mariners, including an ultraviolet spectrometer to study atmospheric properties, an infrared spec-

Figure 12. The Mariner 9 spacecraft was a windmill-shaped spacecraft, like the other Mariner spacecraft that visited Mars before it. The windmill shape was caused by its four solar panels that provided electrical power to the spacecraft. Mariner 9 has the distinction of being the first artificial satellite to orbit another planet. (Courtesy NASA)

trometer to analyze atmospheric composition and dust, and an infrared radiometer to measure temperatures. The earlier Mariners each carried a single camera, but Mariner 9 included two: a 50-mm wide-angle lens (with a resolution of 1 to 3 km [0.6 to 1.8 miles]) and a 500-mm narrow-angle lens (with a resolution of 100 to 300 m [328 to 984 feet]).

After a 167-day flight, Mariner 9 swung into orbit around Mars on 14 November 1971, the first artificial satellite to orbit another planet. Upon its arrival Mars was engulfed in a raging dust storm. Astronomers had first observed the storm from Earth-based telescopes in early September when a yellow cloud was spotted in the mid-southern latitudes. The cloud grew and quickly enveloped the rest of the planet. This dust storm would leave the atmosphere of Mars dusty for most of the time Mariner 9 operated at Mars.

Earth-Like Features

According to Tom Thorpe, Mariner 9 television experiment representative, Mars continued to hide its secrets under the huge dust storm. Mars was featureless for about 1 month after Mariner 9's arrival, until one day four dark spots, the tops of huge mountains, poked through the clouds just north of the equator in the Tharsis region (Figure 13). One of the spots was familiar to astronomers who had discovered it in earlier dust storms. First known as Nix Olympica and later renamed Olympus Mons, it towered far above the other three mountains. Geologists concluded that these were none other than huge volcanoes. The other three volcanoes, also over 20 km (12 miles) high, were named Arsia Mons, Pavonis Mons, and Ascraeus Mons.

After the discovery of the largest known volcano, Mariner 9 went

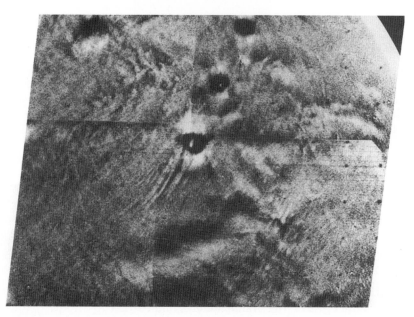

Figure 13. Mosaic of four pictures taken by Mariner 9 of Mars on 13 November 1971 showing the fours dark spots, found later to be the tops of huge volcanoes in the Tharsis region. The mottled appearance of the scene is caused by swirling dust suspended in the atmosphere by a huge global dust storm. The parallel streaks emanating from the southernmost dark spot (Arsia Mons) is a wake in the atmosphere produced by winds driving clouds over the mountain. (Courtesy NASA)

on to make two more major discoveries. The second of these was Valles Marineris, a great chasm that stretches 5,000 km (3,000 miles) along the equator of Mars and in some spots is more than 700 km (420 miles) wide. If it were laid out on Earth, this Grand Canyon of Mars would fill the expanse between Los Angeles and New York City.

Long, dry channels, some with teardrop-shaped islands, were the third major discovery. Resembling river valleys on Earth, Mariner's cameras revealed valleys and channels wandering through the area known as Chryse Planitia. Harold Masursky, the leader of the television team, when questioned about these channels and valleys during a press conference in June 1973, stated that "We are forced to no other conclusion but that we are seeing the effects of water on Mars."

Mariner 9 revealed a planet much different from what the earlier Mariners saw: volcanoes and lava flows, high plateaus, faults, uplifts, depressions, slides, valleys, and tributaries. Geologically, Mars appeared to be a two-part world. The Southern Hemisphere looked ancient and cratered, but the Northern Hemisphere was almost craterless, appearing smoothed by repeated lava flows, yet gorged with canyons, dry river channels, and volcanoes.

Another first for Mars exploration was the decision for the Mariner 9 project to obtain close-up views of Phobos and Deimos, the moons of Mars, when the spacecraft arrived during the dust storm. With only the tops of the volcanoes visible on Mars for almost 2 months, pointing the cameras toward the moons was a phenomenal idea. From Earth-based telescopes Phobos and Deimos appeared as mere pinpoints of light. In Mariner 9's cameras, however, they resembled giant floating potatoes. They remain some of the darkest objects ever photographed in our solar system.

Major contributions from Mariner 9 came from other instruments on board the spacecraft. Winds greater than 200 km/hr (120 miles/hour) were clocked using the infrared spectrometer. Knowing the dynamics of a planetary atmosphere other than our own is extremely valuable for understanding Earth's atmospheric circulation (a problem of great significance to meteorologists) and for understanding the potential effects of pollution on Earth's atmosphere. The infrared radiometer and ultraviolet spectrometer analyzed the atmosphere and found it to contain 90 percent carbon dioxide, though they also detected clouds that contained water vapor.

While scientists were busy interpreting data returned from Mariner 9, preparations were under way for Project Viking. The Vikings consisted of a pair of spacecraft that were each really two spacecraft in one: an orbiter that carried a host of television cameras and remote-sensing instruments for studying the planet and a robot lander for analyzing the surface and to search for life.

After analyzing hundreds of images, thirty-five sites were chosen for closer inspection before the end of the Mariner mission. On 4 June 1972, Mariner 9 began its "extended mission" to complete the mapping and landing site requirements for the upcoming Viking mission. Because there was not as much attitude control gas fuel left as originally anticipated, the extended mission phase had to be divided between requests for landing site images from Viking scientists and those from the Mariner 9 experimenters. Thus, the number of preselected sites was cut to twenty-four. Choosing the best location to land the robots was a difficult chore because of the continuous streams of new data that inundated scientists daily. Some thought was given to landing at the north pole but this was quickly dismissed when team members decided the spacecraft could not operate in the severe temperatures and that it was unlikely any living organisms could survive in such a northern latitude. Viking Lander 1 was targeted for the Chryse Planitia valley north of the 6-km (3.6-mile) -deep Valles Marineris, and Viking Lander 2 for Cydonia.

A peak in the history of American science, Mariner 9 ran out of attitude control gas on 27 October 1972. The tumbling Mariner 9 silently orbits Mars and will remain in orbit for a minimum of 50 years before it enters the Martian atmosphere and disintegrates.

The Soviets took advantage of the same Mars launch opportunity used by Mariner 9. Mars 2 and 3 were launched shortly before Mariner 9, but taking a shorter flight path, Mariner arrived at Mars 2 weeks earlier. Thus, it became the first man-made object to go into orbit around another planet.

As Mars 2 approached Mars, it ejected a capsule that contained a lander. According to V. G. Perminov, the lead designer of Mars and Venus spacecraft at the Lavochkin Design Bureau in the Soviet Union, the computer aboard the Mars 2 capsule suffered a malfunction. This caused the computer to "issue the wrong command to decrease the height of the pericenter of the flyby hyperbola [the aim point for a successful landing]. As a result, the lander entered the Martian atmosphere at a big angle and hit the Martian surface before the parachute system was activated."

Mars 3 also carried a lander attached to the orbiter. Unlike the

Mars 2 lander, after release from the orbiter the Mars 3 lander arrived safely on the surface, becoming the first spacecraft to land successfully on Mars. It touched down at 45°S and 158°W and began to broadcast. But problems started almost immediately, Perminov recalled: "the telephotometer data were transmitted . . . . In 14.5 seconds the signal disappeared. The same thing happened to the second telephotometer." Why had both systems mysteriously stopped within a hundredth of a second of each other? He speculated later on, "I discovered . . . during World war, British radio operators had their transmitters malfunction because of a coronal discharge while working in the deserts of Lebanon during dust storms. On Mars the size of dust particles, the humidity, and the atmospheric pressure are much less, but the wind velocity much higher than in the desert of Lebanon. Perhaps, the coronal discharge was the reason the signal from Mars suddenly disappeared."

Though the landers failed, the orbiters continued to go about their business of collecting data. As with Mariner 9, for a time they could only stare into the billowing dust clouds until the raging storm abated. Before these missions were over, they had probed the atmosphere and surface from orbit and, most important, had accomplished the first soft landing on Mars.

Determined to complete a successful mission to Mars, the Soviets launched Mars 4 and 5 on 21 and 25 July 1973, respectively (Figure 14). These were followed by two orbiter/lander combinations, Mars 6 and 7 on 5 and 9 August 1973 (Figure 15). The Soviets announced that Mars 4 and 5 would continue the exploration begun by Mars 2 and 3.

Mars 4 passed within 2,200 km (1,540 miles) of the rocky surface on 10 February 1974 but failed to go into orbit because its braking engine did not fire. However, during its flyby, photographs were taken and later transmitted to Earth.

Mars 5 maneuvered into orbit on 12 February 1974 and awaited the arrival of its companion craft due 1 month later to provide a communications link with Earth. On 9 March, Mars 7 approached Mars but missed the planet by 1,300 km (780 miles) when an onboard system malfunctioned. The Mars 6 orbiter developed communication problems soon after launch but was able to carry its lander to Mars using commands stored in its on-board computer. On 12 March the Mars 6 lander separated from the orbiter and descended to the surface near 24°S, 19°W. Its radio signals continued

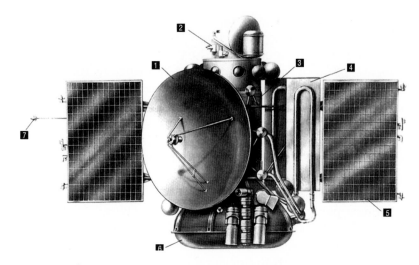

Figure 14. The Soviet Mars 4 and 5 spacecraft. The picture shows (1) the high-gain antenna, (2) radiometer, (3) fuel tank, (4) radiator of the temperature control system, (5) solar panel, (6) instrument module, (7) magnetometer. (Courtesy NASA, from V. B. Perminov, *The Difficult Road to Mars,* NASA Publication 1999-06-251-HQ, 1999)

Figure 15. The Soviet Mars 6 and 7 spacecraft. The aeroshield is the structure at the top of the spacecraft. The aeroshield protected the lander during initial atmosphere entry. (Courtesy NASA, from V. B. Perminov, *The Difficult Road to Mars,* NASA Publication 1999-06-251-HQ, 1999)

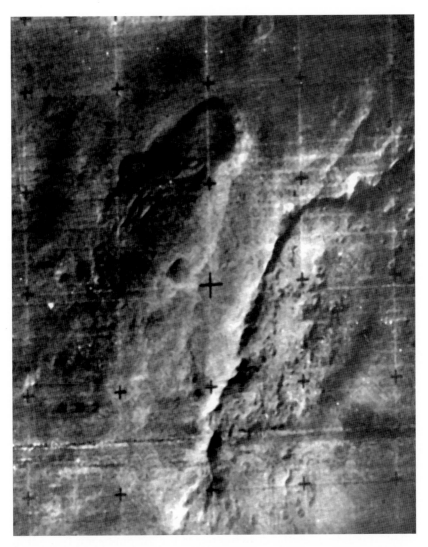

Figure 16. High-resolution image of the surface of Mars obtained by the
Soviet Mars 5 spacecraft. (Courtesy NASA, from V. B. Perminov,
The Difficult Road to Mars, NASA Publication 1999-06-251-HQ, 1999)

for just 2 minutes before the Mars 6 lander crashed. Mars 5 continued its vigil in orbit over Mars, collecting data (Figure 16) and waiting for signals from a lander that never came.

The failures were traced mainly to defects in the electronic components. The minister of electronic industry, A. I. Shokhin, who was responsible for building the electronic components that failed, was asked the direct question by Project M-71, "How do we avoid numerous failures during the next flight?" His answer was simple: new electronic components must be designed and developed. Though the answer was simple, the consequences were not and according to Perminov would eventually "cost almost 10 times the previous cost." Fifteen years would pass before another Soviet mission to the Red Planet.

No Signs of Life?

After Mariner 9, no one really expected large life forms capable of gigantic construction projects on Mars. But in the summer of 1976 when two Viking spacecraft (Figure 17) reached Mars to search for extraterrestrial life, it was not without hope that they would find something.

On 19 June 1976 the first Viking spacecraft went into orbit around Mars. After cruising in space for 10 months, Viking Orbiter 1, as it was called, was 288 million km (173 million miles) from Earth and 1,302 km (781 miles) above the surface of Mars. The higher-resolution images taken by Viking 1 revealed details not visible in earlier photographs taken by Mariner 9. Even after years of intense scrutiny by more than three hundred scientists, the proposed Viking Lander 1 landing site turned out to be on the floor of a deeply carved riverbed, not in the smooth, safe place that the Mariner 9 images implied. It was back to the drawing board for the photo-interpretation teams at Jet Propulsion Laboratory. Two reasons were given for the differences in what Mariner 9 and Viking saw: the much higher-quality Viking cameras and the clearer atmosphere. (There was no dust storm raging when the Vikings arrived.)

It seemed as if everyone had their own idea of what Mars was going to look like when the Vikings arrived. Gerry Soffen, Viking project scientist, recalled: "Everyone was shocked by the craters and geological details revealed by the orbiter's high-resolution cameras. It was during this period that the Viking team went crazy. We had

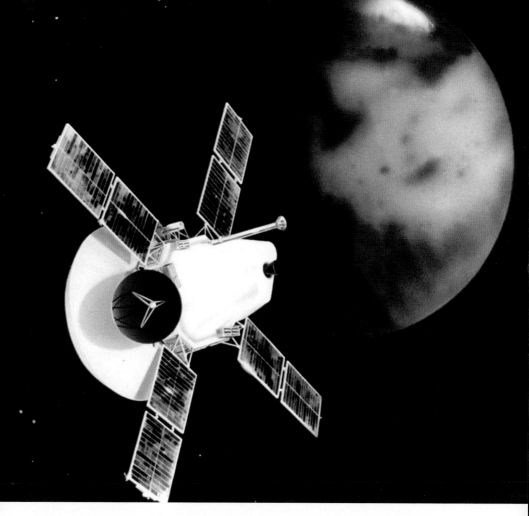

Figure 17. Drawing of the Viking spacecraft in orbit around Mars. The Viking orbiter had the same general layout as the previous windmill-shaped Mariner spacecraft. The Viking lander is folded up inside the capsule at the left of the orbiter. (Courtesy NASA) (76-HC-625)

Viking Orbiter 1 almost ready to detach its lander and we did not know where to land it. We only knew where not to land. I can truly say we never slept during this time."

Sixteen extra days were spent searching for the final landing site, during which time there were many hot debates. Every morning the previous day's photos were hung on the walls for excruciating analysis. It appeared as if a contest was being conducted for the best site. The radar team was the underdog because their data most often indicated that the favored sites were too hilly or filled with boulders. Landing a spacecraft in an unknown area made for some of the most trying times for team members. An acceptable site was finally agreed upon—800 km (480 miles) northwest of the original site, on the edge of Chryse Planitia. Harold Masursky, the leader of the Viking landing site team, explained, "The selected site considered at 47.5°W, 22.4°N represented a compromise between desirable characteristics observed with visual images and those inferred from Earth-based radar." It was the best compromise the team could find between radar roughness and visual smoothness.

On 20 July 1976 commands were uplinked by engineers at Jet Propulsion Laboratory through the giant communication antennas of the Deep Space Network to Viking Orbiter 1 and then to the guidance control and sequencing computer on board the lander to begin the landing sequence. At 5:12 A.M. Pacific Daylight Time on 20 July 1976, the seventh anniversary of a human's first step on the Moon, Viking Lander 1 touched down safely on Chryse Planitia (22°N latitude, 48°W longitude). The touchdown was actually 20 minutes earlier, but because of the great distance to Mars it took that long for Viking's radio signal to arrive.

Once on the surface, the lander went to work immediately, collecting data about Mars and sending the information home. Upon seeing the first image, Viking project manager Jim Martin recalled: "It was all so unbelievable. Here was this lander 200 million miles away, sitting on another planet, and we were on Earth looking at pieces of dirt on its footpad. I felt a little strange. It was such an exhilarating experience to go through nine years of mission planning, and finally that first picture of Mars appears right before your eyes. The first picture that mankind had ever seen from the surface of another planet (Mars). It was so clear; it was as if we were standing right there beside it."

When the second image rolled across the television monitors a desertlike panorama filled with huge boulders and rocks came into view. A gigantic rock, much bigger than the lander, sat a mere 6.08 m (20 feet) away. (It was later named Big Joe.) "The realization of having landed just 20 feet away from destruction was a little unnerving," Martin said. "We could have hit the darn thing and turned the lander upside down and it would have been all over."

From their positions about 1.5 m (5 feet) above the ground, the cameras provided a human's-eye view of the terrain surrounding the craft. Team members speculated that the Martian sky would be deep blue due to the lack of a substantial atmosphere. But when the first color image showed a red Martian sky, scientists assumed that the camera's color tuning was out of adjustment. After some computer enhancement, the sky was made blue and the ground bluish gray. But after more detailed analysis, the flight controllers were convinced that the lander's color controls had been set right initially and the Martian sky was indeed red. How to prove it? They devised a plan to command the camera to photograph the rocket motor that housed a red cable. This was a true color indicator because the spacecraft team knew the cable was red. When the picture came back to Earth, everything was red: the sky, the ground, and the cable. Scientists now think that the reddish tinge was due to the presence of soil particles suspended in the thin air.

Back on Earth, Viking scientists were poring over incoming photographs from Viking Orbiter 1 and studying radar data for safe landing sites for Viking Lander 2. Just as with the first landing, debates and uncertainty reigned as the landing site team tried once again to reach agreement.

There were two landing site locations originally planned for Viking Lander 2. The primary landing site was designated "B-1," an area in Cydonia at 44°N. The backup site, also located at 44°N, was in the Alba region. These sites were chosen because of expected high concentration of water vapor at that latitude. They were a compromise between the desire to get as close as possible to the northern seasonal polar cap deposits while keeping temperature limits consistent with biologically available water.

Even though the site had been selected, there was still disagreement. Based on Mariner 9 photographs, some geologists felt that B-1 was not an acceptable landing site. Again, Viking Orbiter 1 was used in the landing-site search. High-resolution images from Viking Orbiter 1 showed both sites to be unacceptably rough. But by this time the second orbiter was fast approaching Mars with a scheduled orbit insertion of 7 August. Because of the continued disagreement, the original orbit was modified slightly to allow the spacecraft more time to view the original candidate landing sites. Nearly half of the surface of Mars between 40° and 50°N was photographed to locate a more acceptable landing site. For a time there was some consideration given to changing the latitude to the equatorial belt or the Southern Hemisphere. But there was a strong desire to stay where there would be the most water and for a soft, smooth landing site. So the search went on for a region at 44°N containing sand dunes that had migrated from the northern mantled region. After considerable and heated debate, the site was selected in Utopia Planitia at 44°N and 226°W.

The Viking mission was the first time in NASA history that four spacecraft were in operation at once. So that engineers and scientists could concentrate on the descent of the second lander to Mars, Viking Lander 1 was put into a reduced-operations mode.

Separation was successful, and Viking Lander 2 plunged through the atmosphere to Mars. Communications were lost temporarily, but engineering data began trickling in after a command from the Deep Space Network activated the low-gain transmission to relay engineering data.

All was silent at the anticipated moment of touchdown. There was no indication in the engineering data that Viking Lander 2 had landed on Mars. Everyone knew if the lander crashed there would be no signal. Finally, after a ½-minute delay, which seemed like an eternity, the signal was received. Smiles, cheers, and sighs of relief broke out. Viking Lander 2 successfully touched down in Utopia Planitia (48°N latitude, 226°W longitude) on 3 September 1976 at 3:58 P.M. Pacific Daylight Time, ready to go to work.

Early the next morning the first image from the Viking Lander 2 site appeared on Mission Control television monitors. The high-resolution photograph of the lander's footpad was similar to the first photo taken by Viking Lander 1. Subsequent photographs had a peculiar angle to them—the Martian horizon was tilted. Apparently one of the footpads had parked itself on a rock. Instead of sand, everywhere there were rocks of all shapes and sizes resting on a fine-grained soil. Off in the distance low ridges could be seen. Were they sand dunes?

It did not take long for both landers to set about the primary task of searching for life on Mars. The details of this search are discussed in chapter 7. Most scientists believe that Viking did not detect life. But one biology instrument, the labeled release experiment, produced a response that satisfied the original Viking mission criteria for a positive result. Other biology instruments also reacted to the soils, though not in ways expected if life was present. In addition, no trace of organic material was found in the soils. This was thought to be the most telling observation of all. The Viking lander observations combined to suggest to most scientists that soil chemistry produced the positive response in the labeled release experiment and that the soils were sterile. But because Viking could only perform a few simple experiments, these results did not close the question of life on Mars. Instead, they did cause scientists to reconsider their assumptions about the nature and likely location of life and how to search for it.

Post-Viking Blues

The triumph of the Viking mission is even more spectacular when viewed in hindsight of more than two decades of further attempts to reach Mars. The Soviets were the first to attempt a post-Viking return to Mars with the twin spacecraft Phobos 1 and 2 in 1988 (Figure 18). The Phobos project was a multispacecraft mission designed to make comprehensive studies of Mars, Phobos, the Sun, and the interplanetary medium. The Phobos 1 and 2 spacecraft consisted of two orbiters equipped with two types of landers: a long-lived automated lander and a hopper lander. The hopper could move about the surface of Phobos in a series of mechanically induced hops. Phobos 1 was lost on its way to Mars when its high-gain antenna lost lock on Earth and communication ended. Phobos 2 went into orbit around Mars and narrowed its path to within 150 km (90 miles) of Phobos (Figure 19) before unexplainably falling silent.

The United States was the next to try its luck at further exploration of Mars with the launch of Mars Observer in 1992. Mars Observer carried a full load of sophisticated scientific instruments to measure the geochemical and geophysical characteristics of Mars as well as to gather information about the Martian climate. A fuel-line rupture shortly before orbit insertion sent Mars Observer out of control and flying past Mars without being heard from again. The Mars Observer failure reminded NASA that exploring Mars is a very difficult endeavor.

But, undaunted, the United States and Russia held fast to their commitments to explore Mars. In 1996 the largest fleet of spacecraft since the early 1970s was readied to visit Mars. The United States prepared Mars Pathfinder and Mars Global Surveyor, and the Russians made ready Mars '96.

A New View of Mars

Born out of the loss of Mars Observer and based on a vision for a NASA long-term commitment to explore Mars, the Mars Surveyor

Figure 18. Drawing of the Soviet Phobos 2 spacecraft in orbit above Phobos. The landed elements of the mission (a hopper and small lander) can be seen on the surface. High-energy particles stream toward the surface to eject molecules that can be measured by instruments on the spacecraft. (Courtesy NASA, from V. B. Perminov, *The Difficult Road to Mars,* NASA Publication 1999-06-251-HQ, 1999)

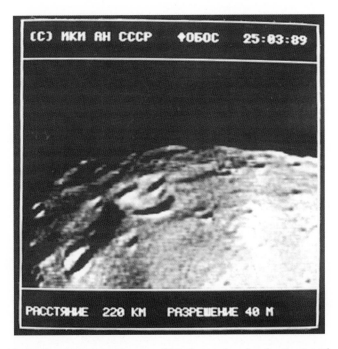

[C] ИКИ АН СССР   ♦ФОБОС   25:03:89

РАССТЯНИЕ   220 КМ   РАЗРЕШЕНИЕ 40 М

Figure 19. High-resolution picture of Phobos taken by the Soviet Phobos 2 spacecraft. The spacecraft was lost soon after this picture was taken. (Courtesy Russian Space Science Institute)

Program was formulated in 1993 and the first mission (Mars Global Surveyor) was launched on 7 November 1996. Mars Global Surveyor (Figure 20) was sent to Mars to recover the science loss as a result of the Mars Observer's failure. It carried many of the same instruments as Mars Observer. Many feel that its arrival at Mars on 11 September 1997 ushered in what has been called the second great era of Mars exploration.

Though Mars Global Surveyor was planned to do its mapping chore from a circular orbit, it was initially placed in an elliptical orbit. Capturing a spacecraft in an elliptical orbit requires much less energy and retro-rocket fuel than does capturing a spacecraft directly into a circular orbit. This means that a spacecraft such as Mars Global Surveyor can be less massive and hence less expensive. But how to get it from the elliptical orbit to the circular one required for mapping? Mission planners devised a scheme to slowly work Mars Global Surveyor into a circular orbit through a process called

Figure 20. In this drawing the Mars Global Surveyor spacecraft keeps a close watch on Mars as it orbits above the planet. (Courtesy NASA)

aerobraking. They would command the spacecraft to dip into the atmosphere at the low point in its orbit to be slowed by the effects of atmospheric drag. There was one hitch to this plan: problems with the locking mechanism on its solar panels left these essential elements of the spacecraft vulnerable to the stress put on them by the drag through the atmosphere. So to avoid destroying the solar panels, Mars Global Surveyor waited nearly 2 years as mission controllers carefully felt their way through the aerobraking process.

As did those of its earlier cousins (i.e., the Mariners and Vikings), observations made by Mars Global Surveyor changed our perception of Mars. For example, Mars Global Surveyor spent more time and in a more elliptical orbit than planned. This carried it closer than planned to the surface each time it dipped into the atmosphere. Though this gave mission controllers fits worrying about the safety of the spacecraft, it carried the magnetometer close enough to the surface to make the first measurements of the details of the planet's magnetic field. Much to the surprise of most scientists, evidence was found of strong remnant magnetism produced by a once-active global magnetic field over the first billion years of Mars's history. High-resolution pictures of the surface revealed small channels that must have been produced recently by running water. Infrared scans of the surface found a large area that might have hosted abundant thermal springs. Mars Global Surveyor also carried a laser altimeter that gave the first accurate three-dimensional view of the surface of Mars. This has been critical in understanding the gravity field as well as what caused many of the Martian landforms.

But the United States desperately wanted to get back to the surface of Mars and resume its exploration for the search for life. It was clear that this could only be done if landing was made much more affordable. Viking had cost about 1 billion dollars in 1976. Knowing that the Viking approach was too expensive, NASA charted a course for a low-cost return to the surface of Mars. Building on what it had learned from Viking and the Soviets combined with new and cheaper technology, it set out to develop Mars Pathfinder. The original concept for Mars Pathfinder was devised by Geoffrey Briggs, the director of NASA's Solar System Exploration, and was based on the Russian semihard lander concept. It was Scott Hubbard at NASA's Ames Research Center who showed how a semihard lander—American style—could work successfully. To handle the tough task of surviving a high-speed landing, he adopted the new "air bag" technology

Figure 21. Mars Pathfinder as seen by Sojourner Rover. The picture is the first one taken of a lander on the surface of Mars by another spacecraft. The tall structure is the camera mounted on its lattice mast. The meteorology mast is to the right. The air bags show prominently in the foreground. (Courtesy NASA) (PIA01121)

from the automobile industry. Then it was up to Tony Spear at Jet Propulsion Laboratory, the Mars Pathfinder project manager, to make it work on Mars. And make it work he did. The result was one of the most spectacularly successful Mars missions ever (Figure 21).

Unlike all other missions sent to Mars, Mars Pathfinder was not designed as a science mission but as a technology demonstration of new and inexpensive systems for entry, descent, and landing on Mars. During the development phases of the mission, almost as an after-thought, a small six-wheeled rover was added to demonstrate rover technology. Named Sojourner, after Sojourner Truth, the Civil War abolitionist, this little rover proved to be a major success and was one of the greatest highlights of the mission.

Mars Pathfinder was launched on 4 December 1996. Unlike the case with Viking, there was no parking in orbit to debate about where to land. Mars Pathfinder went on a direct trajectory to its landing site on Mars. On 4 July 1997 Mars Pathfinder slammed into the Martian surface in a large channel in Ares Vallis and made at least fifteen bounces on its air bags (the first bounce was nearly the height of a four-story building) before coming to rest to radio back "All is A-OK." The Mars Pathfinder control room at Jet Propulsion Laboratory erupted into cheers and tears of joy as the lander retracted its air bags, opened its clamshell-like solar panels, and pointed its high-gain antenna at Earth.

The Mars Pathfinder Lander, named the Sagan Memorial Station after it landed, carried a color stereo imager and an atmospheric structure/meteorology package. After a thorough checkout of the Sagan Memorial Station, controllers at Jet Propulsion Laboratory drove Sojourner off the lander and onto the surface to become the first Martian automobile (Figure 22). Sojourner carried a beer can–sized alpha-proton X-ray spectrometer to nearby rocks to measure their composition. Before completing its mission, it had traversed nearly the length of a football field, dodged between rocks, rolled over several different types of soils measuring the properties of soils and rocks, and taken numerous "dog's-eye view" pictures with its tiny cameras.

Upon landing, the imager on the lander popped up on top of a small boom that gave it a human's eye-level view of the surrounding surface. Pictures from the lander showed a complex surface of ridges and troughs covered by rocks that had been transported to the site by the huge floods that carved the Ares Vallis and from nearby impact craters. Color pictures and alpha-proton X-ray spectrometer measurements of nearby rocks indicated the presence of two closely related types of volcanic rock (see chapter 5). One type, a relatively silicon-rich volcanic rock called andesite, is probably derived from the surrounding volcanic plains, excavated by impact craters. The second type, an iron-rich volcanic rock called basalt, is probably derived kilometers to the south, brought in by the catastrophic floods that carved the Ares Vallis. Several different types of fine-grained materials in the soil were found at the site, as well as evidence for a soil crust similar to that seen at the Viking sites. The weather at this site was also similar to that recorded by Viking 1 nearly 21 years earlier. Mars Pathfinder recorded evidence of dust devils, capturing several in pictures and sensing their effects with the meteorology instruments.

In the fall of 1997, after more than 3 months of operation on the surface, autumn had come to the landing site and the Sun had moved so far south in the sky that it no longer provided enough light to power these little robots. At this point, and having far exceeded their original goals, the Sagan Memorial Station and Sojourner Rover were powered down, their mission complete. Mars Pathfinder had given us another way to get safely to the surface of Mars. Mat-

Figure 22. Newly deployed Sojourner Rover sits at the bottom of Pathfinder's aft off-ramp, waiting commands to start its investigation of the Martian surface. (Courtesy NASA) (PIA01551)

tew Golombek, the Mars Pathfinder project scientist at Jet Propulsion Laboratory, proudly proclaimed that, "The Pathfinder Landing system and its rover are suitable for the exploration of the wide variety of terrains and surface materials on Mars."

Although the United States was enjoying success during this time, the Russians were still plagued by failure. Mars '96 was an ambitious combination orbiter/lander mission built by the Russians. The orbiter carried penetrators and a small lander that were to descend to the surface of Mars. In November 1996 Mars '96 was launched but failed to reach a stable Earth orbit.

Years earlier, noting the difficulty of sending spacecraft to Mars, both the Mariner 9 and Viking teams had joked that according to one of the Jet Propulsion Laboratory's project managers, John Cassani, there is a mythical being, the "Great Galactic Ghoul," that lives between Earth and Mars. This huge monster frequently ate parts from spacecraft on their way to Mars, accounting for the many mysterious failures. It appeared that this mythical being had a taste for Russian spacecraft. As we will see, the Great Galactic Ghoul's appetite was not limited to Russian spacecraft. It also found American spacecraft to its liking and at the next launch opportunity in 1998 was treated to a four-course meal.

A Bitter Lesson: Revenge of the Great Galactic Ghoul

The 1998 Mars launch opportunity was greeted by Mars scientists with excitement and expectation. Five spacecraft, four from the United States and one from Japan, were ready to voyage to Mars to begin the next phase of exploration. Instead of continuing the line of successes started by Mars Global Surveyor and Mars Pathfinder, this opportunity saw the loss of all four spacecraft from the United States and the near loss of the Japanese mission.

In 1998 the United States was aggressively moving ahead with exploration of Mars under the NASA mantra at the time, "faster, better, cheaper." Building on past success and experience, NASA planned to fly two spacecraft, Mars Climate Orbiter and Mars Polar Lander, at a cost scarcely more than that of Mars Pathfinder and less than 10 percent that of Viking. Yet they were full-science missions, containing all the elements of Viking. Mars Climate Orbiter and Mars Polar Lander were built under NASA's Mars Surveyor Program. Mars Polar Lander carried two microprobe/hard landers built under NASA's Deep Space technology program to Mars.

Was it asking too much to fly four spacecraft to Mars at this tiny cost? After all, half of the spacecraft sent to Mars failed and each had cost many times more than those planned in 1998. Exploring Mars is very difficult and requires attention to every detail. Could all the details be covered under such a modest budget? Was the cheap approach tempting the Great Galactic Ghoul? The answer came soon.

In October 1998 a navigation error caused the Mars Climate Orbiter to be inserted into orbit nearly 100 km (60 miles) too low. As a result, Mars Climate Orbiter either burned up in the atmosphere or skipped off the top of the atmosphere into orbit around the Sun. The root cause of the error was reported as a failure to convert English to metric units between flight teams.

Following close behind, Mars Polar Lander, carrying the two Deep Space 2 probes, prepared for its controlled plunge to the surface. The plan was for Mars Polar Lander to jettison the probes shortly before entering the atmosphere and soft land in a manner similar to the Viking landings. In December 1998, shortly before deploying the probes and entering the atmosphere, Mars Polar Lander sent its last signal. It then turned its heat shield toward Mars and radio antenna away from Earth, preparing for entry into the atmosphere. At the time of the turn, all systems were "go." Controllers at the Jet Propulsion Laboratory waited for the signal to come from the surface that the landing had gone well. None came. To the distress of the controllers, none came from the probes either. The lander and the probes had all failed.

Though it is not conclusively known why the lander failed, later analysis found that the small rocket motors designed to lower Mars Polar Lander gently the last 30–50 m (100–150 feet) to the surface did not fire. The Deep Space 2 probes were also silent. Why these probes also failed is a mystery. Perhaps the failure of all three provides a valuable clue. Were their failures in some way connected?

The United States was not the only nation to launch a mission to Mars during the 1998 launch opportunity. The Japanese Nozomi (Hope) mission was launched from Kagoshima Space Center on 3 July 1998 to study the atmosphere of Mars. But trouble started early in the mission. The initial propulsion burn and correction intended to send Nozomi in the proper trajectory to Mars failed due to a sticky valve in the propulsion system. Quick action by mission planners

Figure 23. The Mars Odyssey spacecraft is designed to map the composition of the surface of Mars. (Courtesy NASA)

put the spacecraft on a new trajectory that changed the cruise time from approximately 1 year to 4 years. A fortunate save, and a small price to pay considering the potential return.

Pressing on to Mars

Loss of all four of its spacecraft was a "wake-up call" to NASA and a cause to reevaluate its low-cost approach to exploration. NASA found that its "faster, better, cheaper" approach had been taken too far—the failures were in management and not with the machines. But as with the Apollo 1 fire, NASA's response to this bitter lesson has been to stiffen its resolve, vastly improve its approach, and continue to press on with the exploration of Mars. As President John Kennedy said at the beginning of the Space Program, "We do these things not because they are easy, but because they are hard."

During the next launch opportunity, the exploration of Mars by the United States continued. NASA launched Mars Odyssey (Figure 23), its next mission to Mars and challenger to the Great Galactic Ghoul. On 23 October 2001 Mars Odyssey was placed successfully into orbit around Mars. Mars Odyssey joined its sister spacecraft Mars Global Surveyor, continuing the plan laid out by NASA for the long-term exploration of Mars and cheating the Great Galactic Ghoul of another victim.

Chapter 4

The Interior of Mars

What do we currently know about the planet Mars? To answer this question we will explore the interior, its physical and chemical characteristics, and how they have changed through time. In this exploration we will learn how and why Mars evolved. But as in any journey we should start at the beginning: the origin of Mars and the solar system. Mars is very much a product of the environment in which it was formed and is strongly influenced by processes that operated during its birth. Consequently, understanding how the solar system formed is essential for understanding any planet.

Knowing what is inside a planet is critical to our understanding of its history. By using indirect methods, such as listening to seismic waves or measuring the effects of the variations in mass on the gravity field, we can infer what is inside a body.

On Mars we have only limited direct measurements of what is inside, mainly provided by measurements of bulk properties such as mass, size, mean density, gravity field variations, and moment of inertia. What we have learned from these measurements has been greatly extended through theoretical studies. In spite of the scarcity of data, scientists have managed to piece together enough information, both observational and theoretical, to develop a self-consistent picture of what the interior of Mars is like and how it got that way. This picture of the interior of Mars also has been critical to our understanding of how other planets work.

Origin

A little over 5 billion years ago, Mars was tiny bits of stardust and gas scattered throughout a huge cloud (nebula). These particles came from stars in our galaxy, expelled during their dying stages. Our solar system, including Mars, began to form when these billions and billions of tiny particles, acting under the influence of their own gravity, collapsed inward toward their collective center of gravity. What initiated this inward collapse is not known and is a matter of ongoing debate.

As part of the collapse process, the cloud began to spiral, forming a rotating disk. A body began to form at the center of the disk directly from the collapsing nebula. This body grew so enormous that its internal pressure and temperature were high enough to ignite nuclear reactions, giving birth to a brand new star, our Sun. Its brilliant outpouring of radiation and particles, called the solar wind, made short work of what was left of its parent cloud, quickly sweeping it away from the zone where the planets formed.

As the dust and gas particles in the disk spiraled inward, they collided. Some bounced off one another, but others stuck together to form larger particles. This process is called accretion. Eddies developed in this spiraling disk that became the nurseries for the planets and satellites. As particles were swirled around in these eddies, they accreted other particles; some grew to be planets and satellites. Dust and gas in some of these eddies may have collapsed directly to form massive gaseous planets such as Jupiter and Saturn.

Considering that all the solar system bodies were born from a single swirling cloud of dust and gas, why are some so different from others? The answer to this centers around the Sun's effects on the developing planets and on their evolutionary paths during and after their formation. The Sun controlled the type of materials and where they could condense and accrete. Closest to the Sun, temperature and pressure in the cloud were high and only materials (such as rocky ones) that are stable under those conditions could condense. Farther away from the Sun both temperature and pressure were lower. Materials that are stable under those conditions, such as various types of ice, could condense along with the rocky materials. As a result, the solar system is divided roughly into two zones, the inner solar system where rocky planets and satellites dwell and the outer solar system where ice and gas-rich (volatile-rich) planets and satellites reside.

Other major differences can result from the way a body initially was assembled, in particular if the high-temperature dusty/rocky and the low-temperature gas/icy components mixed as a body accreted. Whether and to what extent mixing occurs is controlled by conditions in the cloud. If rocky and volatile materials come together while the planet forms, then they will most likely react chemically to form new components. In the case of Mars, most scientists think that this is why its density is relatively low compared with the other inner solar system planets. They suggest that the interior of Mars must contain abundant iron oxide produced as primordial water and metallic iron mixed when the planet accreted. The reaction between metallic iron and water not only would produce iron oxide and release hydrogen but also would destroy much of the original inventory of Martian water and metallic iron.

In contrast, Earth is thought to have assembled more heterogeneously, with high-temperature metal components accreting first and volatile components showing up in large amounts only at the end. As a result, Earth's share of metals and water was unmixed and only limited chemical interaction between them could occur. This has left Earth with proportionally more high-density components in its interior than Mars and consequently a higher bulk density.

Thermal History

Regardless of the type of accretion, ultimately all planets formed through collision processes. With debris swirling around in the early solar system, collisions between particles were frequent. As these particles crashed into each other, some stuck together, eventually growing into asteroid or moon-sized bodies. As more particles grew into larger bodies, their collisions became more frequent and violent. Collisions of these bodies with primordial Mars also became more frequent and violent. Toward the end of accretion such collisions were so common that their impacts with Mars released enough heat energy to melt much of the outer portion of the planet.

This original melting set in motion events that soon melted the entire planet. When this melting began, the dense materials that spread throughout the molten layer began to settle, releasing gravitational energy. The gravitational energy was converted directly to heat energy, causing the interior temperature to rise and more melting to occur. This process continued until the entire planet was molten. In the end, the high-density components had sunk to the center, where they formed a metallic core. At the same time, the low-density components floated to the top to form a silicon/aluminum-rich crust. This process, called differentiation, gave Mars an onion-like, stratified interior similar to Earth's. Differentiation was essentially complete within a few hundred million years after the end of accretion.

Differentiation caused the semimolten Martian interior to behave like an enormous churning cauldron. Immediately following this huge thermal event, the interior of Mars began to cool monotonically and solidify. Throughout the rest of its history, the loss of residual heat left over from the differentiation of Mars has had important effects on the planet.

As on Earth, following its initial burst of heat, the churning and chemical processing of the Martian mantle removed much of its heat-producing radioactive isotopes such as uranium, thorium, and potassium, concentrating them in the crust. Most chemical elements consist of more than one isotope, each having different properties (e.g., mass). Some isotopes are radioactive. Radioactive isotopes are unstable and can transform spontaneously into another isotope by emission of subatomic particles. When this happens, heat energy is released and absorbed by the surrounding rock. The slow decay of radioactive isotopes of these elements provides the only source of new heat to slow the cooling of the interior.

The abundance of the radioactive isotopes of these elements dictates how much heat is produced and how dynamic the interior will be as time progresses. The abundance of these isotopes is proportional to the amount of rocky material present. Consequently, for terrestrial planets, size counts. Small, rocky planets, such as Mercury and satellites like the Moon, contain less of these isotopes. Their interiors rapidly expended the heat necessary to drive their evolution. In contrast, because Earth and Venus are relatively large, they have long and dynamic geological histories. Mars is intermediate in size between these two extremes, and it is not surprising that its degree of evolution is intermediate between that of the small planets and the large ones.

The loss of heat from the deep interior of Mars has been achieved by convective overturn in the mantle. This is the most efficient way of transporting heat from the interior of Mars to its surface. To bring heat to the surface where it can be lost rapidly, huge convection cells or plumes of hot rock develop deep inside Mars and then rise to the surface to release their heat; then they cool and dive back down into the interior to be heated once more.

How many of these plumes form in the interior is strongly

influenced by the depth of the heat source and size of the core. Gerry Schubert, a geophysicist at the University of California at Los Angeles, and his colleagues have calculated that the number of plumes decreases and their strength increases as the proportion of heat originating deep in the interior increases and the size of the core decreases. These researchers point out that immediately after core formation, the small molten core of Mars would have provided a strong, deep heat source—perfect conditions for the production of one titanic mantle plume. Indeed, this may have happened. The crustal dichotomy may be the surface expression of a single enormous plume that formed before the planet cooled to a point where multiple plumes came into existence.

With the continued and rapid loss of heat from the interior, more plumes would have developed. Schubert has suggested that, most likely, less than a dozen plumes currently exist in the Martian mantle and that geometrically they are "upwellings in the form of cylindrical-like plumes and downwellings in the form of interconnecting sheets." These plumes feed heat to hot spots in the crust, some of which have given birth to enormous volcanic centers above them. According to Schubert, "The Tharsis and Elysium volcanic provinces are probably the consequence of plume-delivered heat and magma" to the crust. If he is right, this raises a question about the location of the other ten hot spots and their associated volcanic centers. Noting that there are no other major similar volcanic provinces, Schubert has suggested that perhaps the other plumes may have been weaker or the properties of the lithosphere, the structurally strong outer layer of rock, may have caused selection of only one or two of them for prominent surface expression.

There is another plausible, although more complex, cooling scenario to explain the history of Mars. In this scenario it is assumed that the crust of early Mars recycled much like plate tectonics caused changes on Earth. Within several hundred million years, this process stopped and Mars evolved slowly to a tectonic regime like that of the other terrestrial planets, where the crust is rigid and immobile. David Stevenson of the California Institute of Technology pointed out that "If this regime follows one of lithospheric recycling, then the mantle must heat up, because the elimination of heat is less efficient. In other words, the coldest time for the Martian mantle was early in Mars' history, despite the inexorable monotonic decline of radioactive heat source in the mantle and crust."

Bulk Properties

Some of the most valuable information about the interior of a planet comes from a knowledge of its bulk properties (e.g., mass, size, density, composition, and moment of inertia). These properties are used as a basis to predict many other interior characteristics that provide a deeper insight about the nature of the planet and how it evolved.

Before the exploration of Mars, only its mass was known with any precision, derived from effects on the motion of its satellites by the gravity of Mars. But knowing the mass provides little useful information about the planet unless combined with other properties such as its size and shape. A combination of these allows calculation of bulk density and moment of inertia. From bulk density the bulk composition can be estimated, and from moment of inertia it can be determined whether a planet has differentiated into layers.

Beginning with the first spacecraft to orbit Mars, Mariner 9, the size and shape of the planet were measured with enough accuracy to calculate a relatively accurate mean density and moment of inertia. The diameter of Mars was measured using occultation of the spacecraft. In this technique, when an orbiting spacecraft passes behind a planet, its radio signal is blocked. Precisely when and how long the radio signal is lost is measured and used to calculate the planet's diameter. Measured in this way the mean diameter of Mars was found to be 6,778 km (4,067 miles).

The density of Mars, 3.93 g cm^{-3} (0.079 pounds/cubic inch), was calculated from this diameter. Compared with Earth and the other terrestrial planets, this density is low. Earth has a density of 5.52 g cm^{-3} (0.110 pounds/cubic inch), and Venus and Mercury have densities of 5.25 g cm^{-3} (0.105 pounds/cubic inch) and 5.44 g cm^{-3} (0.109 pounds/cubic inch), respectively. Considering that the terrestrial planets are made of materials that accreted at about the same time and in the same general part of the solar system, why should their densities be different?

Some of the differences in density can be attributed to the effects of pressure on the material inside these planets. Earth and Venus each contain more mass than Mars and, consequently, have greater gravity fields that produce higher internal pressures. As a result of these pressures, materials with the same composition deep inside these planets are squeezed into a more compact, denser state than the same materials deep inside Mars.

Pressure effects can be removed theoretically to allow a more direct comparison of the properties of materials inside the planets. When this is done, Mars stands out as being composed of the lowest-density materials. The calculations show that, at zero pressure, the average "uncompressed" density of the materials in Earth's interior is about 4.04 g cm^{-3} (0.081 pounds/cubic inch), and those in Mars are about 3.75 g cm^{-3} (0.0175 pounds/cubic inch). This difference in density is an indicator that the rock inside Mars must be of a different composition than those in the other terrestrial planets.

For Mars, its low density is probably a consequence of the way it formed. During accretion, the mixing of metallic iron with primordial water produced chemical reactions that oxidized the high-density metallic iron to low-density iron oxides in the interior of Mars. For other terrestrial planets with higher mean densities, their metallic iron and water probably did not mix appreciably during accretion. As a result, the interior of Mars contains proportionally more iron oxide than is found in the other terrestrial planets. Much of this material is located in the Martian mantle.

With an accurate measurement of the mass, size, and shape of Mars, its moment of inertia, a mathematically derived estimate of the distribution of mass outward from the central axis, can also be calculated. The moment of inertia is an indicator of whether or not a planet has a layered interior, but its value does not indicate how thick the layers are or their composition. For Mars, a calculation of the moment of inertia using the current values of mass, size, and shape indicates that Mars is layered inside, similar to Earth, with a core, mantle, and crust.

The composition of a planet is also one of the most important and useful of its properties. The best way to determine the composition of a planet is to analyze rocks from it. For such rocks to be most useful, the location where they were derived must be known so that their context can be understood. Currently, we have no such rocks to study. Until we collect rocks from Mars, the next best way of determining the composition is by calculating it theoretically. This can be done with reasonable accuracy based on theoretical considerations, constrained by other better-known bulk properties of Mars, together with extrapolations from the bulk composition of Earth and primitive meteorites. In these calculations, the major adjustable components are generally the concentration of iron oxide in the mantle and the concentration of light elements such as sulfur and oxygen in the

Table 3

Calculated Bulk Composition of Mars Using Three Different Types of Material

Constituents	Composition Similar to Chondritic Meteorites (%)	Composition Calculated from SNC[a] Composition (%)	Terrestrial Mantle and Crust (%)
Mantle and crust			
SiO_2	41.6	44.4	45.1
TiO_2	0.3	0.1	0.2
Al_2O_3	6.4	3.0	4.0
Cr_2O_3	0.6	0.8	0.5
MgO	29.8	30.2	38.3
FeO	15.8	17.9	7.8
MnO	0.15	0.5	0.1
CaO	5.2	2.4	3.5
Na_2O	0.1	0.5	0.3
H_2O	0.001	0.004	——
K (ppm)	59	305	260
Core			
Fe	88.1	77.8	——
Ni	8.0	8.0	——
S	3.5	14.2	——
Calculated relative mass			
Mantle plus crust	81.0	78.3	——
Core	19.0	21.7	——

[a]Shergottites, nakhlites, and chassigny meteorites.

core. Table 3 shows the bulk composition of Mars calculated based on these constraints.

Because we have not yet obtained samples from Mars, these calculated estimates of its bulk composition remain loosely constrained and based on indirect information. Considering this limitation it would be easy to conclude that work on this subject has currently gone as far as it can. Remarkably, this is not the case. Many scientists think that we may already have samples from Mars here on Earth and that these samples are periodically delivered to Earth free of charge. How can we have samples from Mars when we have not sent missions there to bring back samples?

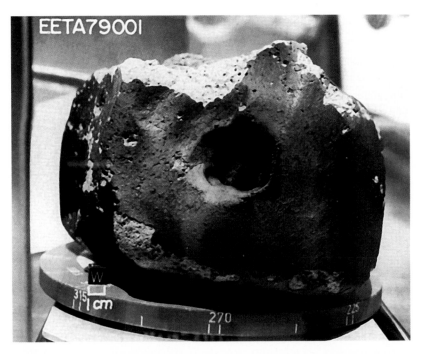

Figure 24. SNC meteorite Elephant Moraine A79001 was found in Antarctica in 1979 at Elephant Moraine. It is a shergottite believed to be fragments of basalt ejected from Mars. (Courtesy NASA/Johnson Space Center)

Rocks from Mars

Before the 1970s most scientists believed that all meteorites were chunks of asteroids or comets. Nearly all meteorites exhibit characteristics consistent with that origin. However, the last quarter of the twentieth century was marked by scientific claims that some meteorites, in particular the SNC meteorites (named after the three major types of igneous meteorites: shergottites, nakhlites, and chassigny), had been blasted off the Martian crust by one or more impacts (Figure 24).

The SNC meteorites have actually been studied for over 150 years. Originally, they were thought to be pieces of igneous rock blasted off a large asteroid, probably Vesta. It was known that in many ways some of these meteorites are similar to terrestrial basaltic rock (common iron-rich, silica-deficient igneous rocks that make up Earth's ocean basins) and others are lhertzolites (course-grained, iron- and magnesium-rich igneous rocks found in Earth's mantle). As more study was done on these rocks it was found that even though these are very different types of rock they share some common attributes, such as their unique oxygen isotope compositions, that indicate that they came from the same parent body.

It was not until the late 1970s that cosmochemists using newly developed high-powered precision laboratory instruments capable of precisely measuring minute amounts of materials began to make discoveries about these peculiar igneous rocks. The discoveries truly shocked the scientific community. Cosmochemists began to see subtle chemical difference between the SNCs, other meteorites, and any known terrestrial rock. Over the past 20 years, information about the chemistry, mineralogy, textures, and ages of the meteorites has accumulated and consistently points to an origin on a relatively large and geologically active planet—most likely Mars.

In 1979 the initial suggestion that the SNC meteorites were from Mars was made by Hap McSween and his colleagues at the University of Tennessee. They argued that based on the late crystallization ages of the SNCs (1.3 billion years to 180 million years ago) they must come from a body large enough to have a long history of igneous activity. Soon after McSween proposed his outrageous hypothesis, Don Bogard and his colleagues at NASA Johnson Space Center in Houston measured the composition of tiny, trapped pockets of air in glasses (solidified molten rock produced by the high-pressure shock associated with formation of impact craters) contained in a shergottite meteorite. These tiny bubbles of gas are thought to be samples of the atmosphere of the body from which these rocks originated. Remarkably Bogard and his colleagues found that the trapped gas is chemically and isotopically a match to the gases in the Martian atmosphere measured by the Viking landers. This chemical and isotopic composition is unique to the Martian atmosphere and provided strong evidence that the SNCs are from Mars. It should be noted here that the relative proportions of isotopes are like fingerprints for the different solar system bodies.

As more information has been collected about these meteorites, the stronger the argument that they come from Mars has become. In particular, a preponderance of geochemical evidence indicates that the SNCs must have come from a moderately large planet. For example, radiometric age dating has shown that some of the SNC meteorites are very young, less than a billion years old, and that they crystallized from magma derived from remelting of mantle rocks. This requires a planet that is large enough to allow melting over much of its history. In addition, the composition of SNC meteorites indicates that the source of these rocks must be a mantle that is depleted in moderately volatile elements such as sodium, sulfur, and potassium, probably a consequence of scavenging of these elements by sulfur-rich melts produced during core formation and global melting. Though other terrestrial planets and our Moon probably started with fewer of these elements, they show no such depletion. The SNCs also have a high concentration of iron oxide, similar to that expected from the bulk properties of Mars and measured in the soils at the Viking landing sites.

Some of these meteorites also show evidence of the type of chemical weathering produced by circulation of liquid water through them, leaving behind weathering by-products such as salts, clays, and carbonate minerals. In addition, the distribution and composition of some of these minerals appear to point to periodic infiltration of small amounts of saline water into the rock like those expected if there were periodic changes in the Martian climate that allowed liquid water on or near the surface.

Initially, a major argument against a Martian origin for the SNCs was the assumption that rock could not survive ejection from Mars without being completely shock melted. The energy required to eject rocks instantaneously from Mars is more than enough to melt them completely. It was argued that if the SNCs were really from Mars, then every one of these meteorites should be only a hunk of glass, a frozen piece of impact melt. Inconsistent with this view, many SNC meteorites are crystalline rocks and show little evidence of shock melting.

Jay Melosh at the University of Arizona and his colleagues provided the answer to the puzzle of how SNCs could be blasted off Mars without being completely melted. They pointed out that when a meteoroid hits a planet it generates a shock wave that radiates outward from a point a little below the surface. This shock wave moves radially from that point and where it hits the surface is reflected back into the subsurface. Near the surface this reflected wave and the radiating wave interfere with each other, in some places canceling the effect. This means that some of the near-surface materials ejected during the crater excavation process are protected from high stress and shock melting. Consequently, no matter how strong the shock wave, some materials only experience low to moderate shock pressures. Melosh and his colleagues had found the key to explain how crystalline rocks on the surface of a planet could be blasted into space and survive without shock melting. Remarkably, this principle holds for all planets and provides hope that we may eventually find meteorites from still other planets on our own planet.

It is also thought that most of the SNC meteorites took a rather circuitous route to Earth after they were blasted off Mars. For dynamical reasons, most materials ejected from Mars should reach Earth within 10 million years of their ejection. But the impact events that blew some of the SNCs into space (probably not all were ejected by the same impact event) occurred at about 190 million years ago. To account for the missing 180 million years, it has been suggested that these SNC meteorites were first ejected into a loose orbit around Mars where they remained for about 180 million years until they collided with a passing asteroid or comet. This collision fragmented the large chunks of the SNC meteorites into smaller pieces and knocked them into orbits that cross Earth's orbit, where they eventually fell as meteorites.

As scientists began to accept the evidence that the SNC meteorites are from Mars, they began to wonder if other types of strange meteorites found were from other planetary bodies. Much to the delight of the believers that SNC meteorites came from Mars, a search of the world's meteorite collections yielded a lunar meteorite. In 1983 Ursula Marvin at the Smithsonian Institution announced the discovery of a meteorite having the same unique characteristics as the samples brought back by Apollo. It should be noted here that without the Apollo samples, such identification would have been nearly impossible. But this discovery provided direct supporting evidence that rocks can be blasted from planetary-sized bodies and find their way to Earth. Can the discovery of samples from Venus and Mercury be far behind?

With mounting evidence that the SNC meteorites are from Mars, scientists are increasingly excited about the potential for using them to solve many of the planet's mysteries. But the use of these samples has two major drawbacks. First, assuming they are from Mars, these meteorites can only be used to deduce general characteristics. More detailed studies require knowing where on Mars these rocks originated. Second, as with lunar meteorites, only a documented sample from Mars would provide conclusive proof of their origin. Though SNC meteorites have provided interesting and potentially very important information about Mars, until samples are obtained conclusions drawn about the planet based on these meteorites will remain speculative.

Up to this point our understanding of the Martian interior has been based on very limited information, mainly from its properties and limited observational data. Consequently, only a general view of the structure and composition of the interior of Mars is possible. Based on our best evidence, as is discussed in the next section, the Martian interior is much like Earth's, composed of concentric layers of decreasingly dense materials that are organized into a core, mantle, and low-density crust. As with our knowledge of Earth's crust, mantle, and core, we know more about the crust of Mars than about its mantle and more about its mantle than its core.

Deep Interior of Mars: The Core and Mantle

The deep interior (i.e., the core and mantle) of Mars is the focus of this section. Though little is known about the deep interior of Mars, some inferences can be drawn about its structure and composition from the properties of Mars, the chemistry of the SNC meteorites, and the recent Mars Global Surveyor magnetometer data. As more observations are made of Mars, it is hoped that its interior properties will become better known.

The Core

Collectively, the data that currently exist about the interior of Mars are only good enough to narrow down a range of characteristics of the core, such as its composition, physical state, and size. Based on bulk properties of Mars alone, its core could be between two plausible extremes in size and composition. At one extreme is a core composed mainly of iron that is rich in oxygen and sulfur, with a density of about 6 g cm^{-3} (372 pounds/cubic foot) and a radius of about 2,200 km (1,320 miles). At the other extreme is a core made of iron-nickel alloy, with a density of about 8 g cm^{-3} (496 pounds/cubic foot) and a radius of about 1,300 km (780 miles). These extremes correspond to a core of between 15 and 30 percent of the total mass of the planet.

The composition of the core, particularly its sulfur content, dictates the physical state of the core and whether it can generate a magnetic field. Planetary magnetic fields are produced by natural dynamos generated by the relative motions between the solid metallic inner core and the molten outer core of a planet. The sulfur content controls whether and how long the core is molten. In the Martian core, a small amount (a few percent) of sulfur would have resulted in the early solidification of an initially molten core and a very short-lived magnetic field. With a core containing a large proportion of sulfur (15 percent), the entire core would have remained molten until the present and a magnetic field would never have been generated. For most sulfur contents between these limits, the core could have generated a magnetic field for part of Martian history before it solidified. However, a core with just the right proportion of sulfur would plausibly still be partially molten, meaning that Mars would have a magnetic field.

Two sources of information help us form our view of the characteristics of the Martian core—the SNC meteorites and the magnetic field data from Mars Global Surveyor. Because of compositional relationships between sulfur in a planet's core and its other chemical components, the composition of SNC meteorites can be used to predict the amount of sulfur in the Martian core. Based on the composition of the SNC meteorites, the core of Mars should be dominantly iron, with 7–8 percent nickel and about 14 percent sulfur. This amount of sulfur is enough for the core to have remained partially molten for nearly a billion years after its formation. During that time it would have been capable of generating a planetary magnetic field.

Consistent with this prediction, Mars Global Surveyor detected what many scientists believe is evidence of an extinct early magnetic field. The magnetometer aboard Mars Global Surveyor measured strong magnetic fields above ancient terrain. No such fields were detected above younger terrain. The simplest explanation for this observation is that Mars had a strong global magnetic field early in its history that is now recorded as a strong remnant magnetic field trapped in its ancient rocks. Judging by the age of the magnetized terrain, the early planetary magnetic field was lost between 3 and 4 billion years ago, about the time that the core should have solidified as predicted from the SNC data. David Stevenson of the California Institute of Technology summarized the implication of this finding: "A core dynamo operated much like Earth's current dynamo, but was probably limited in duration to several hundred million years. The early demise of the dynamo could have risen through a change in the cooling rate of the mantle, even a switch in convective style that led to mantle heating. Mars probably has a liquid, convective outer core and might have a solid inner core like Earth."

What caused Mars to stop generating a magnetic field? Simply, Stevenson pointed out that the planet started out hot but quickly cooled. As the cooling rate declined, the core reached a point where heat flow could be accommodated by conduction alone and convective motion in it stopped. However, it is also possible that the core is richer in sulfur than once thought and that it completely froze early or that early lithosphere recycling ceased, causing the mantle to heat past the temperature where the core could maintain its dynamics.

The Mantle

Although the Martian mantle is thought to be composed of rocks generally similar to those of Earth's mantle, the size of the Martian core has a major influence on characteristics of the mantle, especially its thickness, density, mineralogy, and structure. A large core requires that the Martian mantle be thin and composed of relatively low-density rock; a small core requires a thick mantle made of relatively high-density rock. The range of plausible core sizes and densities allows for the mantle to be from 1,500 to 2,100 km (900 to 1,260 miles) thick, composed of rocks with a corresponding average density of from 3.41 to 3.52 g cm^{-3} (211 to 218 pounds/cubic foot). This is substantially higher than the density of rock found in the mantle of Earth, which is 3.31 g cm^{-3} (205 pounds/cubic foot). The higher density is thought to be a reflection of the proportionally higher concentration of iron oxide in Martian mantle rocks compared with those in the mantle of Earth.

Though a range in density of rock in the mantle is possible, it is thought that the mantle is generally uniform in composition. Depending on its thickness, the Martian mantle may contain three or four successive shells, each having roughly the same composition, although composed of a different type of rocks and minerals. These statements may seem odd at first. Why do we not know exactly how many layers are in the mantle? Equally as odd, why is the composition uniform but each layer is composed of different types of rocks and minerals? The answer to these riddles involves the effects of increasing pressure on rocks and minerals. Minerals have set crystal structures that, when intensely compressed, can be rearranged into new minerals with tighter, denser structures. These tighter arrangements of atoms are new minerals with the same composition but different structures and mineralogic properties. An example of this is carbon: at low pressure it is a soft, black substance called graphite, but at very high pressures its atoms are compressed into a hard, transparent mineral called diamond.

Keep in mind that, depending on the size of the core, the mantle could range from 1,500 to 2,100 km (900 to 1,260 miles) thick. Laboratory studies that squeeze rocks to determine what happens to them at very high pressure have shown that below about 1,600 km (960 miles) depth the Martian mantle should be composed mainly of high-density iron-rich minerals such as majorite and spinel. If the core is larger than 1,600 km (960 miles) in diameter, the innermost shell of the mantle would not exist, nor would these minerals exist in Mars. Above 1,600 km (960 miles) depth in the Martian mantle, majorite and spinel transform into other minerals of identical composition that are 10–12 percent lower in density and greater in volume. Two more zones above the core are expected in the mantle. The mineral transformations in these zones result in about a 10 percent change in volume and density and would be easy to detect if we had seismograms of Mars.

Unlike in Earth, there is little or no evidence that the Martian mantle and crust mix. In Earth, mixing of these two layers occurs and is a product of the subduction of the crust into the mantle associated with plate tectonic processes. In Mars there are no such mechanisms to mix its mantle and crust. Consequently, these two layers probably remained chemically distinct from one another.

On both Earth and Mars, the upper mantle is the source of most magma that feeds surface volcanism. Consequently, the composition of volcanic rock reflects the composition of the upper mantle. The absence of plate tectonics and crustal recycling on Mars has had a considerable effect on the composition of its upper mantle where these melts are commonly derived. As on Earth, when a spot in the upper mantle of Mars melts, the lighter components in the melt readily ascend and the denser ones sink. On Earth plate recycling eventually remixes these components. Without a process to remix the crust and upper mantle of Mars, residual melts in the mantle grow progressively denser. The upper mantle is enriched in high-density components such as iron and magnesium oxides and grows poor in low-density components such as silicon and aluminum oxides. This process is called igneous differentiation and has resulted in the density of the upper mantle of Mars increasing to about the same density as the lower mantle.

A hint to the composition of the magma source is provided by the morphology of volcanic features on the surface, which is partly controlled by the composition of their lava. Generally, the shape of these features on Mars reflects the high-iron, low-silicon composition of the source region. Though most of the iron-rich lava that forms these volcanoes is probably derived from melting of mantle rocks, it is also possible that some could also be derived directly by melting of the primitive mantle. Consequently, the morphology of the volcanoes can suggest a composition of the lava that built them but reveals little about where it came from or the history of its source.

Crust and Lithosphere

Before launching into a discussion of the crust and lithosphere of Mars, we should understand the difference between the two. Often, these terms are incorrectly used interchangeably. One is defined mainly by its composition and the other by its mechanical properties. The crust is the outer, low-density layer of rock compositionally dissimilar to the mantle; the lithosphere is the relatively rigid, structurally coherent, solid layer that includes the crust and the upper part of the mantle.

Because hot rocks are weaker than cool rocks of the same composition, the temperature inside rocky planets is an important factor in controlling the nature of their crust and lithosphere. The loss of heat from Mars has been nonuniform, producing areas of thin lithosphere and dynamic geologic evolution. In places where heat leaks out at a higher rate, the lithosphere is typically thinnest, and the crust generally shows extensive deformation and evidence of invasion by large volumes of magma. In other regions where the flux of heat is lower, the lithosphere is thick. The crust has remained relatively stable and shows little or no signs of volcanism. Most evidence suggests that the Martian crust and lithosphere are stable like those of the smaller planets.

Composition of the Crust

Little is known about the composition of the crust of Mars. We have no direct, comprehensive measurements of its chemistry, only the meager information provided by the instruments on our landers, the SNC meteorites, the bulk properties of the planet, and remote sensing data collected by spacecraft and from Earth.

The information from these sources is still rudimentary and provides only loose constraints on the average composition of the crust. From these data the composition of the crust can only be narrowed to a broad range of possible compositions, any of which are entirely consistent with the available data. For example, within the limits of the data, the Martian crust could be composed of the same materials as the Moon's primitive crust (anorthosite), or as rock from Earth's ocean basins (basalt), or as rock from Earth's continents (granite). The formation of each of these types of rock requires a very different degree of chemical and thermal evolution of the planet. Narrowing the composition to one of these rock types would provide solid evidence about how far the crust has evolved.

In reviewing the available evidence, the SNC meteorites provide powerful, though controversial, hints at the true composition of the crust. The isotopic signatures found in the SNC meteorites are considerably different than those found in rocks for a primitive body, such as the Moon and meteorites, but are similar to those that formed in chemical environments that are more Earth-like than Moon-like. If the SNC meteorites are from Mars, then the Martian crust is composed of rock somewhere in the continuum between basaltic and granitic rock.

Mars Pathfinder and Viking landers provided independent and direct compositional measurements of Martian crustal materials. Unfortunately, because of the limitations of the instruments used to make the measurements, the analyses are not comprehensive, yielding useful but limited information. The Viking landers carried an X-ray fluorescence spectrometer (XRF) that could measure only a few low-density components in the soils. Mars Pathfinder's Sojourner Rover carried an alpha-proton X-ray spectrometer (APXS) that provided an improvement in capabilities by measuring a wide range of elements in both soils and rocks.

The chemical data derived from these instruments suggest that the bulk chemistry of the soil at all three landing sites is similar in many ways (Table 4). This is not surprising. These soils are composed mainly of windblown dust and as a result are derived from all over Mars. Hap McSween, a Mars Pathfinder APXS Team member, and Klaus Kiel of the University of Hawaii, a Viking XRF Team member, studied soil chemistry measurements from all three landers and compared them with terrestrial soils. They concluded that "If the global dust represents a broad average of the Martian surficial or upper crustal composition, the planet's surface geology is dominated by basaltic volcanic rocks and evaporated salts." In short, they suggested that the dust can be traced back to basaltic rocks that weathered in the presence of surface water.

What about the composition of the bedrock at the landing sites? Is it different from the composition of the soils? Is it different at each of the sites? These are difficult questions to answer because of the limitation of the compositional measuring instruments. The composition of the rocks at the Viking lander sites can only be estimated using indirect means (e.g., visual inspection of color pictures from the lander); at the Mars Pathfinder site these indirect means are used in support of APXS measurements. Considering the available data, the rocks at the Viking sites are most likely of basaltic composition (iron, magnesium-rich, and silicon-deficient), consistent with the SNC meteorites and soil composition. More surprising and still controversial, the measurements from the APXS instrument on Mars Pathfinder Sojourner indicate that the elemental composition of the rocks at the Mars Pathfinder landing site is andesitic basalt (i.e., silicon and aluminum-rich basaltic rock) instead of simple basalt. This type of rock contains more light elements and is closer to the average composition of the crust of Earth than to the composition of the ancient crust of the Moon. Rocks with this composition require a geologically active planet to form them. They can be produced either when crust and mantle materials are processed and mixed by plate tectonics or through igneous fractionation and differentiation during multiple episodes of melting and remelting of the crust and upper mantle. Hap McSween argued that, considering the lack of geologic evidence for plate tectonic recycling of the Martian crust, these rocks probably originated through the latter process. He suggested that these rocks are the products of the "early melting of a relatively primitive Martian mantle." The degree of melting and remelt-

Table 4
Chemical Composition of Martian Soils

Sample	Na_2O	MgO	Al_2O_3	SiO_2
Locations where alpha-proton X-ray spectrometer measurements were made by Pathfinder Sojourner[a]				
Near Yogi	3.7	8.1	8.9	46.8
Dark soil next to Yogi	2.7	7.3	8.4	46.5
Scooby Doo	1.9	6.9	8.9	50.3
Lamb Soil	1.5	7.7	8.1	46.8
Mermaid Dune	1.3	7.1	8.2	48.8
Locations where X-ray fluorescence spectrometer measurements were made by Viking 1[b]				
Sandy Flats: fines	——	6.4	8.4	46.0
Sandy Flats: deep fines	——	6.4	7.8	47.0
Rocky Flats: crust	——	7.4	7.2	44.1
Jonesville: fines	——	5.3	7.8	46.6
Rocky Flats: fines	——	6.5	7.7	46.4
Deep Hole 2 tailings: crust	——	7.3	7.3	44.7

[a] Pathfinder data from R. Rieder, T. Economou, H. Wanke, A. Turkevich, J. Crisp, J. Bruckner, G. Dreibus, and H. Y. McSween, The chemical composition of Martian soil and rocks returned by the mobile Alpha Proton X-Ray Spectrometer: Preliminary results from the X-ray mode, *Science* 278:1771–1773, 1997.

[b] Viking data from B. D. Clark, A. K. Baird, R. J. Weldon, D. M. Tsusaki, L. Schnabel, and M. P. Candelaria, Chemical composition of Martian fines, *Journal of Geophysical Research* 87:10059–10067, 1982.

ing of the crust and upper mantle required to produce these rocks implies that Mars has had a dynamic thermal history that far exceeds our Moon's (though less than Earth's).

Remote sensing instruments, both in orbit and Earth-based, have added their own unique information about the composition of the crust. Probably the most important of these came from the thermal emission spectrometer on Mars Global Surveyor. Analysis of spectra from this instrument suggests that there are considerable compositional differences across the surface of Mars. Phil Christensen, of Arizona State University, the thermal emission spectrometer principal investigator, and his teams have analyzed the spectra from this instrument and made a remarkable discovery that "The composition of volcanic materials varies from basalt in the ancient, southern hemisphere highlands to andesite in the younger northern plains."

SiO$_3$	K$_2$O	CaO	TiO$_2$	Fe$_2$O$_3$
6.3	0.2	5.5	1.4	15.6
5.4	0.3	6.3	0.9	18.6
5.2	0.5	7.1	1.1	14.5
6.0	0.2	6.2	1.1	18.8
5.1	0.5	5.8	1.3	18.5
7.5	0.8	6.4	0.7	18.9
7.2	0.9	6.4	0.7	18.5
10.0	0.9	5.9	0.6	18.3
7.2	0.6	6.4	0.7	20.1
6.4	0.7	6.3	0.8	20.3
9.4	0.9	5.6	0.6	18.9

This distribution of rock types is consistent with the measurements made on the surface by the landers. This may suggest a compositional dichotomy that roughly mimics the global age and morphologic dichotomy. Their measurements are also consistent with other spectral and morphological evidence suggesting that the old, cratered terrain (Figure 25), like the lunar highlands, may be composed of more iron-rich rock like that found at the Mars Viking sites (Figure 26).

Collectively, compositional data from a variety of sources have been pieced together to provided a clearer picture of the composition of the crust. But still little is known and many major questions remain to be answered. For example, though there is some evidence that the crust may be made principally of rocks derived from multiple episodes of melting and remelting of primordial materials, it

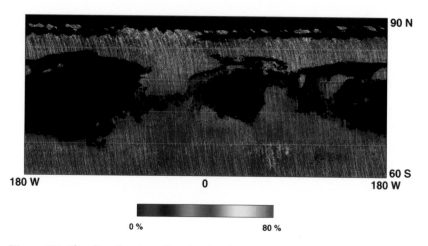

Figure 25. The distribution of rock of andesitic composition on the surface measured by the Mars Global Surveyor thermal emission spectrometer. Purple areas represent 0 percent; red areas represent 80 percent abundance of andesite. The regions where the surface is hidden by wind-blown materials are masked in black. (Courtesy NASA/Jet Propulsion Laboratory/Arizona State University)

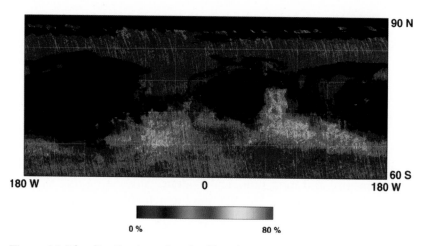

Figure 26. The distribution of rock of basaltic composition on the surface measured by the Mars Global Surveyor thermal emission spectrometer. Purple areas represent 0 percent; red areas represent 80 percent abundance of basalt. The regions where thick dust deposits cover the surface and underlying sand and rock are hidden from view have been masked in black. (Courtesy NASA/Jet Propulsion Laboratory/Arizona State University)

cannot be ruled out that the crust may be composed primarily of primordial materials thinly mantled by younger, more evolved rocks.

Structure of the Crust and Lithosphere

There are a number of methods used to derive information about how the materials are arranged inside a planet (the structure): seismic surveys, variation in the gravity field cause by internal mass variations, analysis of fault patterns on the surface, and the nature and distribution of its volcanism. To a degree each of these approaches has been used on Mars, though some with only scant data available. Still, a picture of the structure of the upper layers of the Martian interior is starting to emerge, suggesting that it is a far more complex and interesting place than originally suspected.

Marsquakes

Probably the best single way to get a clear view of the inside of a planet is through a complete seismic survey. In this type of survey, earthquakes (in the case of Mars, marsquakes), tremors that pass through a planet, are measured and analyzed. Analysis of these shock waves yields three-dimensional slices through a planet in which compositional and structural characteristics are exposed.

Scientists have been listening carefully to the rumbles and murmurs of Earth for many years and have begun to develop a detailed understanding of the nature of Earth's interior. Seismic data have been collected for the Moon and Mars, although these data have severe limitations in both their quality and quantity. Apollo carried seismometers to the moon, but the Moon has little seismic activity. The Viking landers also carried seismometers, but only the one on Viking 2 worked. Its usefulness was severely limited. Its position on the lander made it susceptible to vibrations caused by the wind shaking the lander; also, to collect information about interior structure, seismometers must be placed in an array of three or more to allow triangulation.

In spite of this drawback, the Viking 2 lander seismometer collected unique data that helped to put limits on the type of structure of the crust and lithosphere. During its short time on the surface, this lone seismometer detected at least one good candidate marsquake. The quake was a minute long and had many similarities to earthquakes. To scientists, these similarities suggested similarities in vertical structure and material properties of the crust and lithosphere of Mars and Earth. In contrast, the characteristics of marsquakes and earthquakes are very different from those of moon-quakes, where shocks typically last an hour or longer. "A Moonquake with a similar magnitude and range equal to (the Marsquake and earthquakes) would have generated a seismic signal lasting several hours instead of 1 minute," pointed out the leader of the Viking Seismology Team, Don Anderson of the California Institute of Technology. Anderson and his team suggested that this difference is most likely an indication that, in contrast to the Moon, Mars and Earth have abundant volatile materials, such as water, in their outer layers.

Even with only one seismometer, the Viking Seismology Team managed to extract a good amount of information about Mars. They estimated that the quake was magnitude 3 on the Richter scale and was located about 110 km (66 miles) from the lander. The arrival pattern of the shocks could be interpreted to indicate that the crust is about 16 km (9.6 miles) in thickness at the Viking 2 site in Utopia Planitia. This is generally consistent with the thickness of the crust estimated from gravity measurements (see discussion in the next section) beneath Utopia Planitia (though it is a factor of four or five lower than the average thickness of crust in the Northern Hemisphere). According to Maria Zuber of the Massachusetts Institute of Technology and her colleagues on the Mars Global Surveyor Geophysics Team, the "thin crust beneath Martian impact basin [Utopia in this case] may be the consequence of excavation and mantle rebound associated with the impact process."

The Gravity Field
Of the current data about Mars, irregularities in its gravity field offer the best information about the structure of its crust and lithosphere. Irregularities are produced by variations in mass. If Mars had a uniform structure, its gravity field would show no irregularities. But the Mars gravity field has a rich assortment of irregularities that hint at dynamic internal processes and a considerable history of geologic activity. Some of these variations are due to materials associated with topographic features (i.e., mountains), and others are due to the thickness of the crust or density variation caused by temperature or compositional heterogeneity. Gravity variations are measured by the tiny effect they have on the velocity of orbiting spacecraft and are termed free-air gravity anomalies. Measuring the small velocity

changes is very difficult and requires extremely accurate tracking of the position of the spacecraft.

The mass of the rocks contained (or absent) in topographic features is a major contributor to the variations in the strength of a planet's gravity field and masks differences in the gravity field caused by factors inside the planet. To measure these internal mass variations, the influence of topography on gravity must be removed from the gravity field measurements. In addition to the measurement of the detailed gravity field strength, this requires knowledge of the shape and density of materials that formed the topographic feature. After the effects of topography are subtracted from the total gravity field, Bouguer gravity anomalies, as they are called, emerge from the data to provide a picture of the physical state of the crust and lithosphere.

On Earth, gravity measurements above mountains show little or no anomaly. This is because mountain masses typically float on low-density "root" material that supports the weight of the mountain in a state of gravitational equilibrium. These mountains are said to be isostatically compensated. The low-density mountain roots were first discovered during early gravity surveys of the Himalayan Mountains. Much to the shock of the surveyors, the mass of the mountains had little effect on the gravity field. When the mountain mass was subtracted from the total field, the associated Bouguer anomalies were found to be strongly negative, indicating low-density material beneath.

Recently, Mars Global Surveyor collected global gravity and topography data of unprecedented accuracy and completeness. Except for a few anomalous regions, these new data show that the gravity field of Mars, like Earth's, is also smooth. However, the combination of the global gravity and topography data has been used to produce the first accurate crustal thickness map of Mars (Figure 27). This map shows a distinct north-south trend in crustal thickness, as well as numerous smaller anomalies. In the south, the crust progressively thins from the polar high southern latitudes, where it is about 70–80 km (42–48 miles) thick, northward through the Tharsis province. North of this zone, under the northern plains and the Arabia Terra region of the southern highlands, the crustal structure changes to a zone of nearly uniform thickness where it is about 35–40 km (21–24 miles) thick.

The crust also is thinned beneath all resolvable major impact

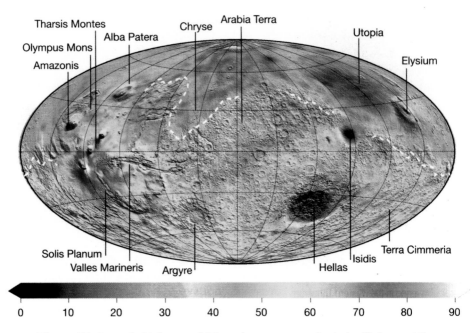

Figure 27. Crustal thickness of Mars shown over a shaded relief map. The dashed line shows the location of the dichotomy between the lowlands in the north and the southern highlands. (Modified from M. T. Zuber, The crust and mantle of Mars, *Nature* [London] 412:220–227, 2001)

basins. This is thought to be due to a combination of excavation and crustal rebound during the impact process. The crust beneath the Tharsis province is thick, representing the accumulation of a massive pile of volcanic materials that are a substantial contributor to the high-standing topography. A thinned crustal region along the central axis of Valles Marineris is similar to that of terrestrial rift zones.

There is no sign of the hypothetical giant impact basin that has been proposed as the cause of the northern lowlands, nor does the location of the transition between the two gravity zones exactly match the boundary of the global dichotomy, as would be expected from the surface geology. Why the subsurface global dichotomy boundary does not match its surface expression is not known.

The thickness of the lithosphere can also be estimated using these new gravity data. The data suggest a general trend: lithosphere thickness generally decreases with increasing surface age, consistent with declining heat flux from the Martian mantle with time. The southern highlands are oldest and have the lowest values of lithosphere thickness (0–20 km [0–12 miles]); the northern plains are youngest and are supported by a thicker lithosphere (about 100 km [60 miles]). The lithosphere of Mars appears to vary in thickness under the major physiographic provinces. Under Alba Patera, the oldest Tharsis volcano, the lithosphere is about 50 km (30 miles) thick, but under Olympus Mons and the other younger Tharsis volcanoes it is 100 km (60 miles) or more.

Lithospheric thickness is thought to reflect the thermal state of the Martian lithosphere in these provinces when they were built and as stress was loaded on them. From the new data, it is clear that thermal conditions in the Martian lithosphere were not uniform over its history. The difference between the thickness of the lithosphere in the ancient southern highlands, northern plains, and Tharsis is a reflection of this and provides a picture of the cooling of Mars over time. Remarkably, the thickness of the lithosphere in the southern highlands indicates that heat flow is considerably less than the predicted global average. If this is correct, then a disproportionally large fraction of heat must have been lost from another part of Mars early in its history. Where?

Maria Zuber and her colleagues on the Mars Global Surveyor Geophysics Team have offered an answer—the northern lowlands. They suggest that the northern lowlands are young and a place "where lithospheric thickness at the earliest times cannot be determined because values in such regions reflect the thermal state at times of subsequent geological activity. High heat flow, distributed approximately over a broad uniform region, would have cooled the mantle and led to lower melt production and to a lower crustal production rate (i.e., thinner crust) later in Martian history. On the basis of crustal structure and relative age, the most plausible locus of high heat flow on Mars is the northern lowlands." This conclusion has far-reaching implications for the early structural style of the crust and lithosphere of Mars. Such high heat flow and wholesale resurfacing of vast regions in the north of Mars requires a different tectonic regime than is evident in the south, prompting speculation about Earth-like behavior of the northern crust early in Mars history.

Plate Tectonics?
Noting the suggestion of the enormous heat flow in the northern lowlands, Sean Solomon of the Carnegie Institute, a leading Mars geophysicist, offered a provocative explanation—ancient Martian

plate tectonics. Building on the work of Zuber, he has speculated that "the northern lowlands was a location of high heat loss from the interior early in Martian history, probably due to a period of vigorous convection and possible plate recycling inside of Mars." Solomon cited the 1994 proposal of geophysicist Norm Sleep, of Stanford University, that earlier in Martian history crustal formation was achieved by a Martian analog of plate tectonics and that the crust of the Northern Hemisphere was formed during the final stages of plate tectonics on Mars.

Adding to the suspicion that the early history of Mars had its own brand of plate tectonics, the southern highlands show broad stripes of alternating magnetism (Figure 28) similar to the magnetic striping observed at spreading centers on Earth. Although these magnetic anomalies may be an indication of a failed attempt at crustal spreading in the southern highlands, it is the northern lowlands crust that shows the effects of being at the center of the most recent plate tectonic activity. The crust beneath the northern lowlands has distinctive crustal thickness variations, but the crust within the zone of magnetic strips is not distinct from other parts of the southern highlands crust. Zuber noted that in the northern lowlands "The 40-km crustal contour corresponds closely to Sleep's proposed limits of late-stage mobile plates. On Earth, passive upwelling of warm mantle material produces a crust 6-km-thick. A 40-km-thick northern lowlands crust could be the product of crustal spreading if the mantle potential temperature were somewhat greater than that typical of modern mid-ocean ridges." She pointed out that this is consistent with early Martian plate tectonics, although it far from proves the hypothesis.

Plate tectonics has important implications on what happens in the interior of a planet, especially with regard to the magnetic field. David Stevenson and Francis Nimmo at California Institute of Technology have calculated the thermal effects on the interior of Mars and the early demise of plate tectonics. They found that the magnetic field and plate tectonics on Mars could have been intimately linked, and that the end of plate tectonics could have hastened the death of the planetary magnetic field.

Small-Scale Structures: Fractures and Wrinkles in the Crust
As on other terrestrial planets, fractures and small wrinkles are common in the crust of Mars. Each of these structures provides a record

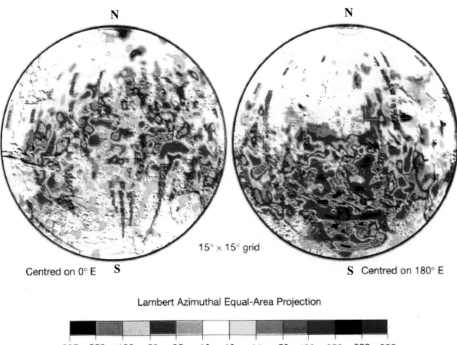

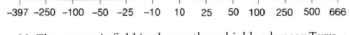

-397 -250 -100 -50 -25 -10 10 25 50 100 250 500 666

Figure 28. The magnetic field in the southern highlands near Terra Cimmeria and Terra Sirenum regions. The magnetic bands are oriented approximately east-west and average about 100 km (60 miles) wide and 600 km (360 miles) long. The blue and red colors represent magnetic fields in the Martian crust that alternately point in opposite directions, giving these stripes similarity to patterns seen in Earth's magnetic field at the midoceanic ridges. (Courtesy NASA/Goddard Space Flight Center) (PIA02008)

of the stress and a suggestion of the nature of the geologic event that produced it. Commonly on Earth structures form patterns that are a product of plate tectonic movement (see discussion in the next section). On Mars structural patterns are similar to those seen on other planets with lithospheres composed of a single rigid plate.

On Mars the most common types of individual structures are fault grabens and wrinkle ridges. Grabens are long, straight fault troughs, produced when the crust is pulled apart (Figure 29). Grabens on Mars range from a few kilometers wide and several hundreds of kilometers long to huge structures 100 km (60 miles) wide and more

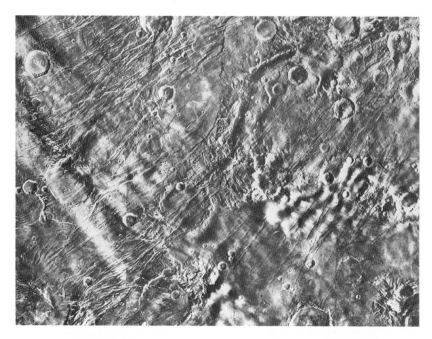

Figure 29. Grabens in the Thaumasia Fossae region. The grabens in this region are generally less than 5 km (3 miles) wide. The scene is about 800 km (480 miles) across. (Courtesy NASA, Viking) (57A04–13)

than 1,000 km (600 miles) long. Sets of these structures are prominent in and around Tharsis, where the effect of loading due to its enormous weight produced strong tensional stresses. In contrast, wrinkle ridges are linear to arcuate asymmetric crinkled ridges formed by compression in the crust (Figure 30). Most are gentle topographic rises tens of kilometers wide, topped by a narrow crenulation or wrinkle. These structures were first studied on the Moon and Mercury. They are found on nearly all types of terrain on Mars, especially on the smooth plains concentric to Tharsis. In addition, only a few isolated faults on Mars show horizontal displacement like that common to faults produced by compression stresses associated with plate tectonics on Earth. The few faults on Mars that show horizontal displacement have developed radially to Tharsis and like many of the wrinkle ridges are probably the result of the stress field produced by its weight.

The size of these structures and their geometric patterns provide

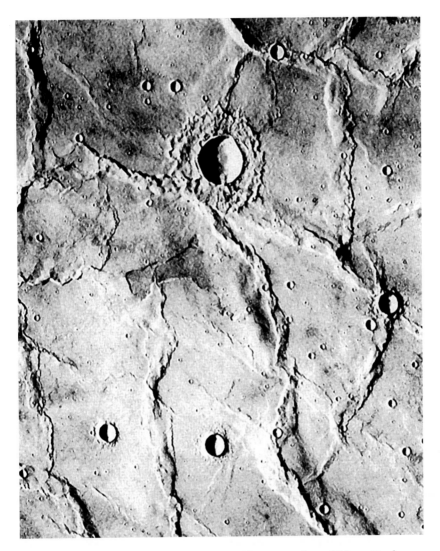

Figure 30. Wrinkle ridge in the Hesperia Planum region of Mars. Each ridge includes a broad rise, a superposed hill or ridge, and a crenulation. The scene is 107 km (64 miles) across. (Courtesy NASA, Viking) (418S39)

a measurement of the strength and thickness of the lithosphere. Based on the analysis of these structures and their patterns, the lithosphere around some of these huge volcanoes is thought to range from about 20 km (12 miles) to greater than 50 km (30 miles) thick when these volcanoes formed. Analysis of faults concentric to Olympus Mons predicts a lithosphere that may be as much as 200 km (120 miles) thick, in agreement with estimates based on gravity.

Structural Style
The structural style of a planet is how its related structural elements (e.g., faults, folds, and other features that indicate deformation) go together into a distinctive system. For example, the style of deformation of the crust and lithosphere of Mars is very different from that of Earth. We know that the structure of Earth's lithosphere is dominated by plate tectonics, but the lithosphere of Mars is a single immobile plate. These two structural styles reflect fundamental differences in the way planets operate.

Earth's lithosphere is broken into a complex mosaic of huge slabs or plates of rock that move with respect to one another. These plates ride on the asthenosphere, the semifluid layer in the upper mantle, and are driven by thermal energy from the interior. The relative movement of these plates creates unique patterns of faults and other structures. On the leading edges of plates where they collide, either a huge arc-shaped trough is pushed down if one of the plates rides over the other or a tall mountain range is pushed up if the plates merely crash into each other. On the sides of plates where they slide past one another, great lateral faults develop, such as the San Andreas fault in California. On the trailing edges of plates, the crust and lithosphere are pulled apart, producing gaping rift zones such as the one in East Africa and the mid-Atlantic. Unique types and patterns of faults and folds are associated with each of these boundaries. These structures and patterns reflect the stress field produced by the particular type of plate interaction.

Earth's internal heat escapes mainly along the plate margins. Consequently, melting of the upper mantle and crust is most common in these areas. But there are also spots in the middle of the plates where melting and volcanism occur. In these places, convection cells transport heat from the deep interior upward to produce "hot spots" and melting in the upper mantle. On Earth volcanism is vigorous at the hot spot but generally short-lived. Movement of the plates shifts the lithosphere above these hot spots before prolonged volcanism can occur in any one spot.

In contrast, the Martian crust and lithosphere is a single rigid, immobile shell. It appears to have been that way since the end of the period of heavy bombardment. Volcanic and tectonic features on Mars, no matter their age, are like those found on other planets with lithospheres composed of a single, immobile plate, such as the Moon and Mercury. Consequently, the structural style of Mars is more like these planets than like Earth.

Because of its structural style, there are no plate margins where heat can be readily released from Mars. It can only lose large amounts of heat in a few places (i.e., hot spots) where convection plumes in the mantle transport heat to both the crust and lithosphere from below. Because the Martian crust and lithosphere remain locked in place and convection cells appear to move very little, hot spots in the Martian crust remain nearly stationary. Consequently, long-lived convection plumes deliver heat to the same places until they die out, resulting in the continued, long-term growth of volcanoes to enormous sizes.

After the solidification of the crust and lithosphere, Mars has seen the development of only two major geologic structures and several smaller ones. As in plate tectonics on Earth, these are a direct result of the structural style of the Martian crust and lithosphere and are discussed individually later in this chapter.

Igneous Activity in the Martian Crust

There is considerable evidence that, as on Earth, melting of the interior of Mars has been common and that the Martian crust and lithosphere have been pervasively invaded by magma. Some of these magmas have spilled out on the surface as volcanic materials. Over its geologic history, Mars has erupted an equivalent of enough volcanic material to completely cover its surface to a depth of about 1.5 km (0.9 mile). Although this may seem like a huge amount of volcanic material, most magma solidifies in the interior and never reaches the surface. On Earth, about ten times more igneous material remains in the subsurface than is erupted. The amount may be even larger for Mars and is consistent with the supposition that the Martian mantle has undergone considerable melting and remelting, causing deep igneous differentiation.

Combined with general constraints provided by the bulk properties of the planet, our knowledge of surface volcanism can be used to infer igneous conditions in the interior, as well as environmental conditions on the surface when volcanism occurred.

Martian Volcanology
The volcanic landforms on Mars typically resemble volcanic landforms on Earth, although there are important and subtle differences. The similarities are produced because both planets are similar in many ways. But Earth and Mars also have some important differences that affect volcanic processes and volcanism; most are environmental factors, such as the relative strength of their gravity fields and their heat budgets. These factors affect almost every process that controls production, cooling, storage, ascent, and eruption of magma.

The most obvious difference between volcanoes on Earth and Mars is the enormous size of some Martian volcanoes. This is partly due to the rigid lithosphere of Mars compared with the mobile lithosphere of Earth.

Nearly as important as lithospheric structure, the weaker Martian gravity field plays a role in making Martian volcanism different than Earth's. In particular, the gravity field has important effects on the buoyancy of magma and its ability to ascend. The buoyancy of magma depends on the strength of the gravity field and density of the magma. On Earth the gravity field is strong and magma ascends rapidly in the crust, but on Mars the gravity field is weaker, and for magma with the same density, buoyancy forces are smaller and magma ascends more slowly. As a consequence, only the most massive of magma bodies in Mars can remain molten long enough to reach the surface and produce volcanism, and small magma bodies probably solidify before ascending high enough to give birth to surface volcanism. Because magma bodies capable of producing surface volcanism must be massive, they commonly pour out enormous volumes of lava onto the surface.

Magma composition affects viscosity and how freely lava can be erupted and flow. Lava rich in iron is runny compared with lava high in silicon. Magma composition also affects buoyancy because it controls magma density. Some estimates of the composition of the Martian lava suggest that it may contain as much as 25 percent iron oxide. Remarkably, the high iron content acts to narrow the range of depths where melts can be generated and still reach the surface. Such a high concentration of iron requires that these magmas originate from less than 300–400 km (180–240 miles) depth. Below that depth such high-iron magmas are compressed to a greater density than the surrounding rock. As a result, if high-iron magma forms deeper than about 400 km (240 miles) in Mars, it would be so dense that it would sink instead of rise, never getting close enough to the surface to erupt.

On Mars buoyancy and crustal structure combine to build Martian volcanoes that reach staggering heights (not just enormous sizes). In the case of Olympus Mons, the magma was buoyant enough for it to ascend about 25 km (15 miles) to the top of this huge volcano from a magma chamber 3 km (1.8 miles) above its base at the level of the surrounding terrain. The magma chamber is up in the body of the volcano. This also happens in some larger terrestrial volcanoes. This occurs when the mass of the mountain above the magma weighs enough to compress its rock to a substantially greater density than the magma. This and the hydrostatic pressure created by the mountain's weight drives the magma upward into the body of the mountain and eventually to the summit.

The ascending magma can stall at barriers in the subsurface before it comes close to the surface. These barriers are in zones where the density contrast between the magma and the surrounding rock is low or in places where the strength characteristics of the rock changes, as on Earth at the asthenosphere/lithosphere boundary. In these cases, magma may still find its way to the surface through fractures that act as magma conduits, called dikes. Dikes are produced when the surrounding rock is pried apart by pressure from growing magma bodies. Most magma carries enormous amounts of dissolved gases and when these gases are liberated they are capable of producing huge pressures.

The style of volcanic eruption (e.g., explosive or effusive) is strongly influenced by the amount and behavior of the volatile materials contained in the magma, such as water or carbon dioxide. Because volatile solubility is partly dependent on pressure, decreased pressure during ascent allows bubbles of gas to form and grow. If there is enough gas, the magma will erupt into fragments carried in a gas stream, resulting in explosive eruptions. Because the Martian gravity is less than Earth's, gas bubbles begin to form and magma fragmentation occurs deeper in Mars than in Earth. As a result, relatively more highly fragmented magma is expected to be produced on Mars.

The thin Martian atmosphere also plays a role in Martian volcanology. Gas released from Martian magma by the fragmentation process rapidly expands into the Martian atmosphere. Because of the comparatively lower pressure of the Martian atmosphere, eruption velocity of the gas is greater on Mars than on Earth. These higher eruption velocities eject fragments of magma, called pyroclasts, farther on Mars than on Earth (also compounded by the low gravity of Mars). When these fragments are hot and molten, they land on the surface and coalesce into lava flows with long run-out distances. When these fragments are cool and solidify as they erupt, the greater eruption velocities on Mars produce widely dispersed, poorly consolidated ash deposits. These deposits form broader, lower-relief edifices compared with those of similar origin on Earth. This mechanism may explain why there is an apparent lack of steep cinder cone volcanoes on Mars.

Explosive volcanism is not always caused by high concentrations of gases dissolved in the magma. On Earth explosive volcanism can be caused when magma encounters groundwater, and huge volumes of steam are generated. This rapidly expanding steam produces enormous pressure within the magma chamber. Breaching of this pressurized magma chamber can have catastrophic consequences. One of the most spectacular examples of this type of explosive eruption that occurred in historical times was in 1883 at Krakatoa, a tiny volcanically active island off Sumatra. When seawater seeped into its magma chamber, a huge steam-driven explosion instantaneously blasted this 800-m (2,608-feet) -high volcanic island into a crater 300 m (978 feet) deep. On Mars there is abundant evidence of a global subsurface water/ice layer. Considering the extensive volcanic history of Mars, the Krakatoa type of explosive eruptions must have been common in some parts of Mars. Even so, there seems to be a scarcity of low-profile volcanoes produced by explosive volcanism, called maar volcanoes. The apparent scarcity of maars may be an observation effect and a result of misidentification due to their close resemblance to impact craters.

If for some reason magma fragmentation does not occur, then the magma will erupt effusively, generally forming relatively thick, long lava flows. The Martian atmosphere also plays a role in controlling how far these lava flows run before they solidify. The extent of lava flows on Mars is strongly influenced by the low cooling efficiency of lava on the surface as well as the higher eruption volumes and rates. The low cooling efficiency of Martian lava flows is partly a product of the inefficiency of the thin Martian atmosphere to carry away heat from the surfaces of these flows. Observations confirm the expectation for longer flows on Mars compared with terrestrial ones. Single lava flows on Mars can contain over 100 km^3 (21 cubic miles) of solidified lava and extend for several hundred kilometers.

In spite of the presence of plentiful volcanic features on Mars and considering the environmental factors such as low gravity and the potential that magmas could get trapped in subsurface barriers, it is a wonder that there is volcanism at all on Mars. Ron Greeley at Arizona State University and his colleagues have done extensive research on Martian volcanology and suggested that these factors "imply systematically greater eruption rates and individual eruption volumes than on Earth." The higher eruption rates and greater volumes have substantial effects on the size and shape of Martian volcanoes. Greeley argued that "Mars' large volcanic edifices, voluminous flows, and inferred pyroclastic deposits on Mars are consistent with predictions of large magma chambers and dikes, eruption volumes and rates, and vigorous explosive activity."

Most volcanic landforms on Mars have morphologic characteristics that are consistent with construction from material of basaltic composition, volcanic materials of high iron and magnesium content and low silicon and aluminum content. In contrast, some volcanoes on Mars, in particular those interpreted to be giant stratovolcanoes (which have steeper flank slopes composed of short, thick lava flows), resemble terrestrial volcanoes made of andesites, a volcanic rock higher in silicon and aluminum content. Consequently, it is tempting to conclude that this is further evidence for extensive processing of the Martian crust. But such a conclusion may not be justified. Ron Greeley pointed out that "because of Mars' unique environment, there is no need to invoke evolved compositions (andesite) to explain explosive features, although large Martian magma chambers could promote the generation of small volumes of more evolved magma via fractional cyrstallization." Consequently, even though morphologically different volcanoes have developed side by side on Mars, this does not necessarily mean that they are different in composition. All of the volcanic mountains could be constructed of the same materials, just erupted under different conditions.

Remarkably, even though proportionally less magma can reach the surface of Mars compared with that of Earth, the most spectacular volcanoes in the solar system are on Mars. This is partly due to differences in the structural style of their crusts and lithospheres and to environmental differences on both planets.

The Volcanic History of Mars

A wide range of types, ages, and sizes of volcanoes are found on Mars. Through painstaking analysis, researchers have pieced together a general view of the long and varied volcanic history of Mars, gaining insight into its flux and interior thermal conditions.

These researchers, in particular David Scott and Kenneth Tanaka of the U.S. Geological Survey, have theorized that very early in its history, as the crust solidified, large volumes of lava poured out on the surface of Mars to form vast lava plains. Because of the destructive effects of early heavy bombardment, remnants of these lava plains have not been preserved.

As the period of heavy bombardment came to an end at about 3.9 billion years ago, massive floods of lava blanketed the older heavily cratered lava plains. These younger floods produced thick lava plains that buried all but the rims of the largest of the preexisting craters. In a few places, lava flow features, such as flow scarps, have survived on these younger lava plains to confirm their origin.

As the interior cooled, this early intense phase of volcanism wound down and the rate of lava plain formation dropped. During that time, interior melting was not as widespread and the transport of heat from the deep interior became restricted to only a few strong mantle plumes. As a result, volcanism became restricted mainly to locations above these plumes.

Piles of volcanic materials began to build up over the locations of these plumes and some of the large central volcanoes began to form. Low-profile highland volcanoes, called paterae, such as Tyrrhena Patera, were the first of the large central-vent volcanoes to be built. They are thought to have grown as a result of explosive eruption of ash produced by volatile-rich and/or silica-rich magma. During that time, the floor of Hellas and Hesperia Planum were flooded by fluid lava that solidified into lava plains. Volcanism also began in the Tharsis region, with the eruption of fluid lava centered at Alba Patera.

For the next billion years, extensive volcanism produced vast lava plains across Mars. The oldest of these is at least 3.2 billion years old and includes the ridged plains in Lunae Planum, Syrtis Planum, Noachis-Isidis Planum, and the Noachis region. About 200 million years after those plains formed, extensive volcanism produced the lava plains in the Amazonis, Syrtis, Elysium, and Chryse regions. The center of volcanism on Mars switched to the Elysium region, where the Elysium volcanoes developed for the next 500 million years. Near the end of major volcanism in Elysium, between 2.0 and 2.5 billion years ago, Acidalia Planitia and the Utopia region were covered by extensive lava flows. These eruptions marked the end of widespread volcanic activity on Mars.

After about 2.0 billion years ago, volcanism became restricted mainly to the Tharsis and Valles Marineris regions. Most of the topography of the enormous volcanic mountains was constructed during that time. The most recent eruptions in these areas were near the giant volcanoes on the northwest flank of Tharsis and along faults in Valles Marineris. Some of the eruptions in Valles Marineris are so young that they could be less than a million years old.

In general, the intensity of volcanic activity on Mars has steadily waned and become more localized. The decline was accompanied by changes in volcanic style, with fewer explosive eruptions occurring later. In some areas, such as the Tharsis and Valles Marineris regions, volcanism has been remarkably long lived (at least 2.5 to 3.0 billion years and perhaps longer).

In the next section, we will take a tour of the most important and spectacular structural and volcanic features that have developed in the crust and lithosphere. Each of these features tells its own story of the evolution of Mars.

A Tourist Guide to the Martian Crust

By now it should be clear that the crust and lithosphere of Mars are not just bland, uniform layers of rock. For the most part, the crust and lithosphere are what they are because of nonuniform loss of heat from the Martian interior. As is discussed later in this section, there are a few cases where major crustal features are not the result of any interior process; instead they result from the formation of huge impact craters. Some features are global in scale; others are smaller but still exert their influence regionally. Together these features make Mars a particularly interesting place and provide a deeper insight into its nature and history.

In this section, imagine that you are a cosmic tourist on a holiday, yearning to see all the points of interest in the Martian crust. You are in luck. A special tour is being offered just for you in this section. So, all aboard, fasten your seat belts, and hold on: we will start our tour with the largest feature of them all—the crustal dichotomy—and progress to smaller and smaller features.

Crustal Dichotomy

As with the ocean basins and continents of Earth, the topography and geology of Mars can be divided into two major provinces (see Figure 4). The origins of these provinces on each planet are most likely due to very different causes. Earth's crustal dichotomy is defined by the stark differences in composition and elevation between the ocean basins and the continents and is a product of plate tectonic processes. The origin of the Mars dichotomy is unknown, but must be a result of a global process.

The northern one-third of Mars is surfaced by sparsely cratered lowland plains. The southern two-thirds of the surface of Mars is composed of heavily cratered highlands that stand an average of about 3 km (1.8 miles) higher than the sparsely cratered lowland plains. Except where obscured by younger volcanic materials, the boundary between these regions is commonly marked by a pronounced cliff 1 km (0.6 mile) high or by a complex of eroded fractures (Figure 31). Fracturing and faulting along this boundary appears to have been active well past the time of heavy bombardment, suggesting that the dichotomy continued to form for a time afterward.

The heavily cratered terrain in the south is surfaced mainly by basaltic rock; in the north it is blanketed with rock of andesitic composition. In places, numerous "ghost" and obscured craters are partially buried by these sparsely cratered plains (Figure 32). These partially buried craters exhibit population densities that are essentially identical to that of the crater populations in the southern highlands, suggesting that the surface buried by the sparsely cratered plains is as old as the southern highlands.

The materials that have buried the old craters in the north vary in thickness within the northern lowlands. In many places, partially buried craters with a diameter of only a few kilometers across show through the fill on these plains. It seems likely that the fill materials consist of sediment transported from the southern highlands early

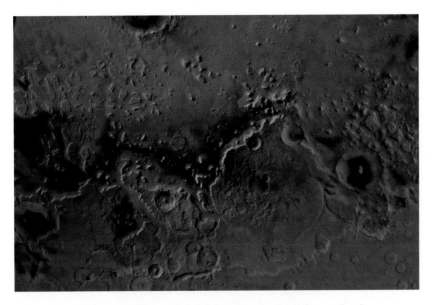

Figure 31. A kilometer (0.6-mile) -high scarp commonly marks the boundary between the sparsely cratered plains to the north and the ancient highlands to the south. The boundary in the Ismenius Lacus region, shown here, has been eroded to a belt of dissected terrain, called fretted terrain. This type of terrain is thought to form as a result of sapping and erosion along lines of structural weakness (i.e., faults). The scene is about 1,110 km (660 miles) across. (Courtesy NASA/U.S. Geological Survey) (PIA00420)

in the history of Mars when the climate permitted water to flow on the surface. Such massive erosion is consistent with other observations that suggest that Mars had an early mild climate (see chapter 6).

Most geologists who study Mars had assumed that a dichotomy in subsurface crustal structure would be associated with the dichotomy in topography and terrain type. However, this does not seem to be the case. Maria Zuber noted this in her analysis of Martian gravity data and concluded, "the geologic expression of the dichotomy boundary is not a fundamental feature of Mars' internal structure." But other geophysicists and geologists disagree and argue that the north-south variations in crustal thickness need not be an exact match with the dichotomy boundary to be the surface expression of a fundamental feature of the crustal structure of Mars.

There have been many hypotheses about the origin of the dichotomy. Some scientists have speculated that there may have been

Figure 32. In places the northern plains show a high-density population of buried craters. These craters appear to be buried by a thin blanket either of sedimentary material deposited from the water that flooded the region or from the wind, or of volcanic materials. (Courtesy NASA/Jet Propulsion Laboratory/Malin Space Science Systems, Incorporated)

an initial asymmetrical distribution of materials in the crust. Others suggest that the northern lowlands were the site of a huge ancient impact event (or multiple overlapping smaller impact events). Other scientists prefer the hypothesis that the dichotomy was formed by forces driven from inside. They attribute it to the effects of convective overturn in the Martian mantle that produces either a region of thin crust or a spreading center such as that seen in midocean ridges on Earth where new crust is formed (see earlier in this section). Recently, it has also been suggested that the geological expression of the dichotomy may be due in part to the piling up of thick ejecta deposits excavated from the Hellas Basin. Though ejecta can account for some of the high-standing terrain in the south, it does not explain all of it. It also does not account for the geologic differences between the hemispheres, the presence of the scarp, or the age of most of the southern highlands, which predate the Hellas impact.

Each of these hypotheses has its supporting evidence and drawbacks. George McGill has studied this region extensively and has suggested that the weight of evidence favors an origin that in some way must involve convection in the mantle. He suggests that "it solves the volume problem better than impact, it is consistent with the geophysical data as impact, and it can be explained using models of internal dynamics similar to models under consideration for Earth."

However, according to Maria Zuber, to produce the observed dichotomy through mantle convection, "a long-wavelength pattern of heat loss with upwelling in one hemisphere and downwelling in the other" must occur. For this to happen the Martian mantle must be layered in viscosity such that the upper mantle is at least one hundred times less viscous than the lower mantle. She adds, "Such a layer could correspond to a Martian asthenosphere, as occurs on Earth beneath the lithosphere. As the viscosity of the lower layer increases relative to the upper layer, deformation becomes more efficient at longer wavelengths." Long-wavelength convective patterns cool the mantle and core more efficiently then short-wavelength convection. Such rapid cooling could contribute to the demise of the core dynamo and magnetic field.

Tharsis: King of Volcano/Tectonic Centers

Tharsis is truly the king of volcano/tectonic complexes. It is the largest complex in our solar system. Only slightly smaller than the crustal dichotomy, Tharsis is a huge plateau that rises about 7 km (4.2 miles) above the average elevation of Mars and extends for about 8,000 km (4,800 miles) across the western equatorial region (Figure 33). It is so large that its mass may even have affected the orientation of the axis of rotation of Mars, possibly causing climatic changes.

It straddles the crustal dichotomy boundary on the western side of Mars. Tharsis consists of two broad rises, a northern and a larger southern rise. The southern rise is superimposed on the highlands and the northern rise is superimposed on the lowlands.

It is the home of spectacular volcanoes and fault complexes. Tharsis contains twelve enormous volcanoes as well as vast lava plains and numerous small volcanic edifices. All twelve of these volcanoes are larger than 90 km (54 miles) across. Of these, five are truly mammoth in size. Four are giant shield volcanoes (Olympus Mons,

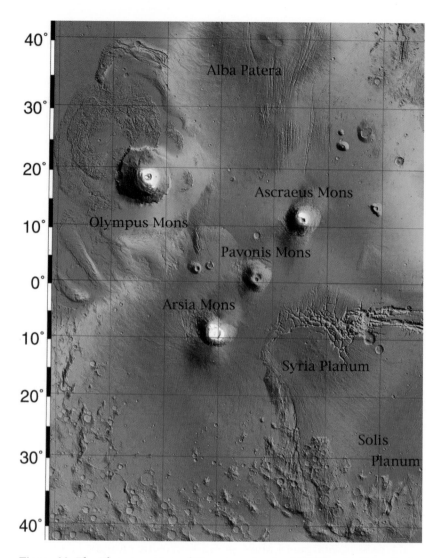

Figure 33. The Tharsis region of Mars is a plateau 8,000 km (4,800 miles) across and 7 km (4.2 miles) high. This altimetry map shows the giant volcanoes of Tharsis as well as the western part of Valles Marineris. Blue is lowest and red and white are the highest regions. (Courtesy NASA/ Jet Propulsion Laboratory/Goddard Space Flight Center)

Pavonis Mons, Arsia Mons, and Ascraeus Mons) that are all over 19 km (11.4 miles) high and 400 km (240 miles) across. Pavonis Mons, Arsia Mons, and Ascraeus Mons are found in a line at the crest of the southern rise. Olympus Mons is located on the lowland, off the western edge of the Tharsis rise. The fifth of these enormous mountains is Alba Patera, a low, broad volcano that has no terrestrial counterpart. Each of these volcanoes is topped by a large complex summit caldera.

Olympus Mons is the largest volcano of its type in our solar system. It has a broad, low, inverted shieldlike profile (Figure 34) that is typical of shield volcanoes on Earth. It is about 550 km (330 miles) across and over 20 km (12 miles) high. The flanks of Olympus Mons are composed of numerous interwoven lava flows. Some of these flows spill onto the surrounding plains and are buried by flows from other nearby volcanoes. The dimension and form of the flows on the flanks of Olympus Mons suggest basaltic composition.

The base of Olympus Mons is marked by a cliff over 1 km (0.6 mile) high. The origin of this cliff is unknown. Some scientists think that it could be a sea cliff cut by an ancient ocean; others suggest

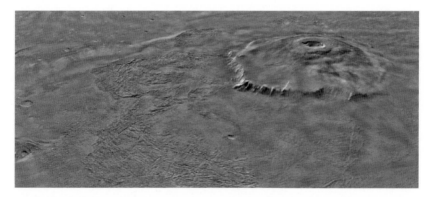

Figure 34. A composite of images and topographic data showing an oblique view of Olympus Mons and its aureole deposit. This huge shield volcano is located to the west of the main Tharsis rise. A kilometer (0.6-mile) -high scarp or cliff surrounds the bases of the main edifice of the volcano. The aureole deposit, the rough hilly terrain in the lower left, may be a huge landslide produced by collapse of the flank of the volcano. Such slides commonly are found around the Hawaiian Islands, where they can be traced across the ocean floor for several hundred kilometers. (Courtesy NASA) (PIA02805)

that it is a fault scarp, a cliff formed by vertical movement on a fault produced by uplift of the mountain.

An enormous fan-shaped deposit extends northwest from the base of the main edifice of Olympus Mons. This is one of the most unique and mysterious features of Tharsis. The kilometer (0.6-mile) -thick deposit is composed of a rugged complex of ridges. Like the cliff, its origin is also unknown, though some scientists suggest that it may be an enormous landslide formed by collapse and run-out of part of the main edifice of Olympus Mons. Other scientists suggest that it is an erosional feature caused by wearing away of soft volcanic ash flows that were deposited as a result of explosive volcanism in the area.

The oldest, broadest, and most unique of the five giant volcanoes, Alba Patera, is about 1,600 km (960 miles) across and 6 km (3.6 miles) high. It is located on the northern rise of Tharsis. Its flank slope is a mere 1°, five times lower than the slopes on the shield volcanoes. It is surrounded by a band of fractures that cut most of the lava flows on its flanks (Figure 35). This relationship indicates that the last major phase of volcanism was over before the regional stress pattern that formed the fractures was imposed. Alba Patera has two discrete calderas. The northernmost of these is partly flooded by lava flows originating from the southern one, indicating that the southern caldera was active more recently than the northern one.

There are seven smaller volcanic mountains on Tharsis. These mountains range from about 90 to 200 km (54 to 120 miles) across and represent a different type of volcano, called tholi (Figure 36). They typically have steeper slopes than do the shield volcanoes and are also topped by calderas that are much wider than is common for such large volcanoes. The extraordinary size of these calderas and their steep slopes are probably an indication that these mountains have been partially buried in thick deposits of younger volcanic materials.

Detailed mapping of the individual volcanic units in the Tharsis region has resulted in the understanding of the sequence of development of volcanoes and their surrounding lava plains. This information provides insight into how Tharsis formed and its evolution. Through this mapping it has been found that Tharsis developed in six major episodes that deformed the rocks in three major centers of uplift. Accompanying these episodes were numerous episodes of volcanism. Most took place in the first half of Martian history, with limited activity continuing into more recent history.

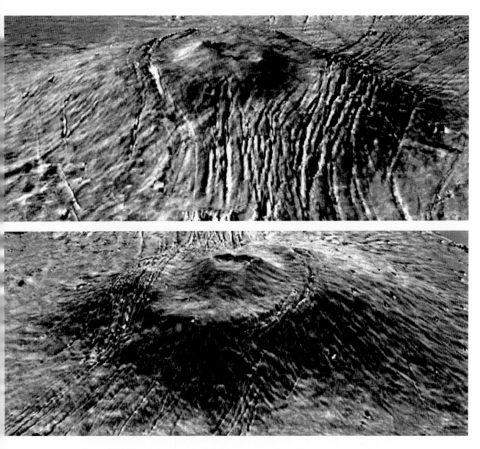

Figure 35. Two views of Alba Patera (from the north [*top*] and the south [*bottom*]) constructed from topographic data collected by the Mars orbiter laser altimeter aboard Mars Global Surveyor draped over a Viking image mosaic. These pictures show the relationship between fault location and topography on and surrounding the Alba volcano. The topography of Alba is exaggerated vertically by a factor of 10 in this picture. (Courtesy NASA) (PIA02803)

The most intense episodes produced spectacular systems of fractures and wrinkle ridges around Tharsis. All of these structures, no matter their age, show an amazing consistency in their orientation. Nearly all fractures are oriented radially to Tharsis. Most of these are grabens, produced by the crust around Tharsis being pulled apart as it was stretching during the episodes of deformation. Wrinkle ridges are found concentric to Tharsis, mainly in the ridged plains to the east. The orientation of these structures suggests that

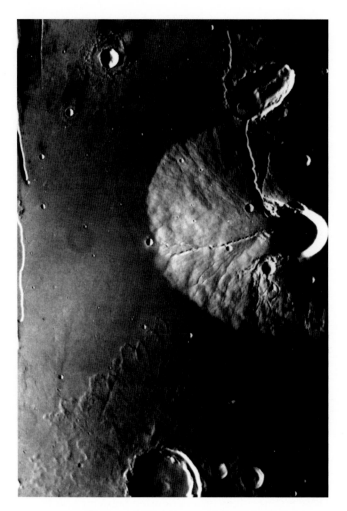

Figure 36. Ceraunius Tholus, a volcano 120 km (72 miles) across, is one of the smaller of the giant volcanic mountains located on the Tharsis Plateau. Numerous fine radial channels are cut into the steep flanks of this volcano. A channel 2 km (1.2 miles) wide originates at the summit caldera and extends down to the base of the volcano. The origin of this channel is unknown, but it could have been carved by volatile-rich volcanic materials that formed density currents after they were erupted from the summit caldera. (Courtesy NASA, Viking) (516A24)

compressional stresses developed around Tharsis, a result of the region being pushed down and outward by the tremendous weight of the topographic rise.

The consistency in orientation of these structures suggests a remarkably stable stress system throughout the development of Tharsis. To some scientists this suggests that the stress system must have resulted from the loading of the lithosphere by the weight of Tharsis.

From the time of its discovery, a debate has raged about how the enormous weight of Tharsis is supported. On Earth, such giant topography must be dynamically supported either by uplift from beneath or by isostatic compensation where it floats on a large root of low-density rock. Earth's lithosphere is far too weak to support any enormous load for long periods. For Mars, none of these mechanisms, by themselves, completely fits the constraints offered by the observational data.

Attempts to explain Tharsis by isostatic compensation have failed, although gravity data suggest that the southern region of Tharsis may be underlain by a crustal root and at least partially isostatically compensated. Gravity data indicate that crustal roots are not found under other regions of Tharsis.

Some scientists have offered alternative hypotheses: either heat from below drives dynamic uplift of the region or Tharsis is mainly a huge pile of volcanic materials held up by a strong lithosphere. Evidence has tended to point more toward the latter as the dominant mechanism. Gravity data indicate that the lithosphere under the Tharsis rise is generally about 150 km (90 miles) thick and grows to about 200 km (120 miles) thick under Olympus Mons. Such spatial variability in thickness suggests that the topography of the region could be supported mainly by the strength of the thick lithosphere and need not be lifted by dynamical forces.

Although Tharsis is thought to be a product of dynamic processes that operate inside Mars, other more controversial hypotheses have been proposed. Though most scientists think that Tharsis formed because it is situated over a large convecting plume of hot mantle, it has also been proposed that the development of Tharsis is related to the formation of the Hellas Basin, which sits roughly antipodal to Tharsis. In this hypothesis, heating and fracturing in the Tharsis region are the result of dissipation of shock energy focused on the opposite side of Mars from the basin. Alternatively, it has been proposed that Tharsis resulted from compressional arching of the lithosphere produced by global cooling and shrinking. This hypothesis requires high stresses to be transferred over great distances and end up focused in one spot. Neither of these models is particularly consistent with current data and as a result they are not widely accepted.

The Grandest Grand Canyon: Valles Marineris

Valles Marineris is an immense canyon system that stretches from western Tharsis for nearly 5,000 km (3,000 miles) across the equatorial region of Mars (Figure 37). In places, it is over 700 km (420 miles) wide and 8–10 km (4.8–6 miles) deep. It makes the Grand Canyon of Arizona (450 km [270 miles] long, 30 km [18 miles] wide, and 1.6 km [1 mile] deep) look like a tiny trench. Because this system of canyons is so large, its presence can be detected using large Earth-based telescopes. As a result, parts of Valles Marineris were named by early astronomers before the exploration of Mars by spacecraft and still bear those names.

The Valles Marineris canyon system is divided into three major regions. The westernmost region of this system is composed of a network of interconnecting short, narrow canyons. The central and main part of the system is composed of multiple, parallel canyons, chains of craters, and down-dropped fault blocks extending for nearly 2,400 km (1,440 miles). The eastern region is composed of a series of irregular depressions. These merge with several collapsed regions that form chaotic terrain farther to the east and are the source of the outflow channels. Each of these regions has it own unique character and records the work of different geologic processes.

Western Valles Marineris is a network of interconnected troughs or canyons, called Noctis Labyrinthus. Individual canyons are relatively short and narrow. Their intersecting pattern and simple trough shapes suggest that these canyons are probably down-dropped fault blocks, formed when western Tharsis was uplifted. Such uplift would have bowed up the crust in this region, stretching and pulling it apart. Some of the valleys in Noctis Labyrinthus also show evidence of enlargement by sapping. This suggests that groundwater was present in the subsurface in this region even after it had been uplifted.

Central Valles Marineris broadens, deepens, and breaks up into several large, parallel canyon segments. East of Noctis Labyrinthus,

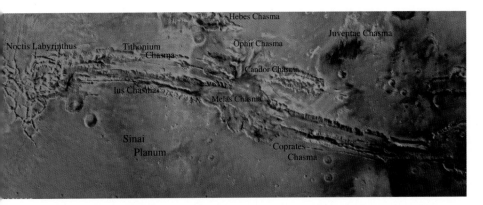

Figure 37. Valles Marineris, the great Martian canyon system, stretches for 5,000 km (3,000 miles) across the equatorial region of the planet. The scene shows the canyon system, extending from Noctis Labyrinthus, the complex system of grabens to the west, to the chaotic terrain to the east that was the source of water for the huge outflow channels that empty into Chryse Planitia. Valles Marineris may have formed from a combination of structural uplift, erosion, and collapse. (Courtesy NASA/U.S. Geological Survey) (PIA00422)

it becomes a broad trough that splits into two huge canyons: Tithonium Chasma and Ius Chasma. The northernmost of these canyons, Tithonium Chasma, narrows eastward and eventually becomes a line of closed depressions. Lines of closed depressions are thought to be indicators that the crust has been pulled apart by the same uplift that formed Noctis Labyrinthus. Ius Chasma, southernmost of these canyons, is composed of two parallel troughs separated in its eastern region by a long, narrow ridge. In places, these canyons are as much as 6 km (3.6 miles) deep. Farther east, central Valles Marineris opens up into Melas, Candor, and Ophir Chasmata, a partially connected canyon system nearly 700 km (420 miles) wide. To the east of these canyons and connecting with Melas Chasma are several smooth-floored, parallel troughs that form the Coprates system of canyons. Three solitary canyons, Echus Chasma, Hebes Chasma, and Juventae Chasma, have formed north of these canyons. Echus Chasma and Juventae Chasma are the source of outflow channels that drain along the east side of Lunae Planum into Chryse Planitia. Hebes Chasma, the central of these three canyons, is a completely enclosed depression and shows no emerging outflow channels.

The walls and floor of the canyons of central Valles Marineris have developed features that record the events of a long and dynamic geologic history. Some of these features were produced by volcanism, and some were produced by the deposition of sediments. Some appear to have required standing water on the surface to form, but others appear to have required only groundwater. Other features are ancient, but some may have formed recently.

Typically, the canyon walls in this central section of Valles Marineris, especially in Ius and Coprates Chasmata, are sculpted into spurs, gullies, and smooth talus slopes. The slopes of the upper parts of these walls are steep and are typically cut into subparallel ribs; in the lower parts of the walls, where the slopes are lower, the walls are cut into sharp-crested, downward-branching spurs. Smooth talus slopes, as well as the scars of large landslides, are common. The spur and gully topography are absent on the slopes of the landslide scars. This suggests that either the spurs and gullies developed at an earlier time when climate conditions were different than after the landslides developed, or the landslides are so recent that spurs and gullies have not had time to develop.

Landslides are particularly well developed in Ius, Ophir, and Hebes Chasmata in central Valles Marineris (Figure 38). They are most common on the north side of the canyons, where their scars typically form curved, smooth, talus-covered reentrants in the canyon walls. Some of these landslide scars are 5 km (3 miles) high and have produced deposits that contain nearly 1,000 km³ (216 cubic miles) of material. The largest of these have flowed out across the canyon floor for nearly 100 km (60 miles). This is much farther than would be expected for a dry landslide on Earth and may indicate that their movement was aided by water. The water could have been either groundwater in the slide materials or water in lakes that may have been in the canyon floors. Alternatively, some scientists think that the great distance traveled by these landslides was simply an effect of the reduced Martian gravity and that these slides are dry-rock avalanches.

The wall rock shows horizontal layering in most parts of the canyon system (Figure 39). The layers, thought to be the exposed edges of layered rock that makes up the surrounding plateaus, are particularly well exposed in central Valles Marineris. Michael Malin of Malin Space Science Systems, Incorporated, has found that "images acquired at 5 to 10 meter/pixel resolution show layering is ubiq-

Figure 38. In this oblique view of Ophir Chasma constructed from topographic data and a Viking image mosaic, landslides are shown that originate from the far wall of the canyon. These slides have a blocky upper layer, which is probably disrupted caprock, and a finely striated lower layer, which is probably mostly debris from the old highly fractured megaregolith beneath. Similar striations are found on some large terrestrial landslides. The scene is a composite of Viking images and topography combined. Ophir Chasma is about 100 km (60 miles) across. (Courtesy NASA/U.S. Geological Survey)

uitous within Vallis Marineris. Layering is seen, where bedrock is exposed throughout the entire depth of the canyon, in places several kilometers below the plateau." This suggests that the plateau Valles Marineris cuts through is composed of layered rock that was deposited, layer upon layer, to a depth of at least several kilometers. What are the layers made of, where did the material come from, and how was it deposited? We do not have the answers to these questions, though some scientists think that these layers are predominately volcanic flows with interlayered deposits of ancient regolith, volcanic ash, and sediments. Others think that these layers represent the accumulation of wind- or water-lain materials deposited within a huge ancient basin. It will take more detailed work and additional types of data (compositional data) to answer these questions.

Figure 39. Layers are clearly visible in this picture of the walls of
Tithonium Chasma/Ius Chasma section of Valles Marineris. These layers
may be part of a thick sequence of volcanic or sedimentary rock that
forms the surrounding ridged plains. The scene is about 5 km (3 miles)
across. (Courtesy NASA/Jet Propulsion Laboratory/Malin Space Science
Systems, Incorporated) (PIA00806)

Numerous tributary canyons cut the walls along the south side of
Ius Chasma (Figure 40). Most of these canyons are less than 100 km
(60 miles) long and 10 km (6 miles) wide. They tend to be nearly the
same width and depth (about 1 km [0.6 mile] deep) throughout their
course and typically merge with the main canyon at the level of the
floor. The heads of these tributary canyons have blunt amphitheater
shapes. Canyon segments are straight and controlled by faults. Their
shapes suggest that they formed by sapping processes. The consistency of depth of the valley heads is probably an indicator of the
depth of the aquifers that provided the water throughout the area.

Young faults are observed running along the walls of the canyons.

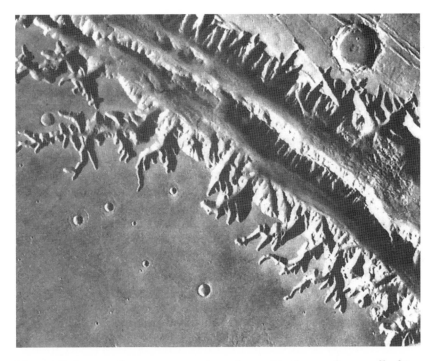

Figure 40. Large tributary canyons have formed in the southern wall of Ius Chasma. These canyons are thought to have formed by sapping processes caused by the action of groundwater escaping from the subsurface in this region. The scene is 300 km (180 miles) wide. (Courtesy NASA, Viking) (645A59)

In places, these faults cut the lower spurs to form triangular facets at the base of the north wall of the canyon. On Earth this type of relationship is common along mountain ranges that are currently being uplifted. Though recent uplift is not expected in this area, this relationship probably indicates the relative youth of these faults. In other places, these faults may have been the source of some of the youngest volcanism on Mars. Barbel Lucchitta of the U.S. Geological Survey has found dark ash deposits along one of these young fault lines that overlie young landslide deposits. The evidence suggests an age of, at most, only several million years.

The floors of these canyons contain remarkable sedimentary deposits. The most remarkable of these are the thick strata found on the floors of Melas, Candor, and Ophir Chasmata (Figure 41). These layered deposits are so thick that in some places they nearly reach the height of the canyon walls. They have been eroded to benches, free-standing tablelands, and mesas. In some places the cliffs around these features terminate in scarps nearly 6 km (4.2 miles) high. This is within only 0.5 km (0.3 mile) of the rim of the surrounding plateau. Many scientists believe that the horizontal layering and lateral continuity of individual layers could only result from deposition of sediment in a low-energy environment. They point out that such an environment is typical of standing water in lakes or small seas. Scientists suggest that the water may have come from the groundwater system, though a candidate source for the enormous amount of sediment required to build these deposits has yet to be identified.

In places there are thinner, irregular-shaped deposits superimposed on some of the older deposits in the canyon floor. Some of these irregular deposits are light-colored and smooth, and others have lobate fronts, mottled albedo patterns, and rough-textured surfaces. Some are associated with pits that resemble volcanic craters and may be ash deposits. Other deposits, particularly those in eastern and central Candor Chasma, appear to be alluvium (material shed from canyon walls). These deposits are dissected by numerous cracks that may be shrinkage fractures resulting from desiccation of wet sediments.

Dunes are found throughout the canyon. Some of the dunes are dark. Outcrops of dark rock, most likely mafic volcanic rock, inside the canyons appear to be the source of this material.

The easternmost part of Valles Marineris consists of two enormous, irregular-shaped depressions, Ganges Chasma in the north and to the south a broader depression that contains Eos Chasma and Capri Chasma. Eos Chasma and Capri Chasma lack the consistent east-west trend of the western canyons. This part of the canyon system is the source of the huge outflow channels that empty into the Chryse Basin (see the section on Outflow Channels in chapter 5).

The floors of these canyons are generally rubbly and grade eastward into chaotic terrain. The chaotic terrain contains jumbled blocks of crustal materials that are typically several hundred meters to several kilometers across. This terrain appears to be crustal materials that had its support removed from beneath it. Because the outflow channels emerge from this terrain full-blown, most scientists think that this terrain formed as a result of the catastrophic release of groundwater and the subsequent collapse of the terrain. They suggest that the groundwater in this region was under tremen-

Figure 41. Thick deposits found in Valles Marineris commonly show layering. In this picture several thousand meters/feet of horizontal layered rock is exposed in buttes and mesas on the floor of western Candor Chasma. These layers might be rocks formed by deposition from water, wind, or volcanism long after Candor opened up or they might be the same materials as exposed in the walls of the chasma. The scene is about 6 km (3.6 miles) across. (Courtesy NASA/Jet Propulsion Laboratory/ Malin Space Science Systems, Incorporated) (PIA01459)

dous pressure, an effect of the uplift of the Tharsis Plateau to the west. In addition, James Dohm and Victor Baker, both at the University of Arizona, have found evidence for an ancient drainage basin in this region, which could have infiltrated and enhanced the local aquifer.

The largest of these chaotic areas, Aureum Chaos and Hydroates Chaos, connect with two large channels, Simud and Tiu Valleys. These channels emerge from the terrain and drain northward into eastern Chryse Basin. Farther east, several smaller patches of chaotic terrain in Margautifer Chaos, Aram Chaos, and Iami Chaos connect with Ares Vallis and are the sources of channels that drain northward into Chryse Planitia.

How and why the Valles Marineris canyon system formed is still one of the greatest mysteries of Mars. Since its discovery, its origin has been a subject of considerable debate. Most scientists think that several different mechanisms were important in producing this enormous canyon system. There is considerable evidence that tectonism played an important role, as well as depositional and erosional processes. The thick-layered deposits and the chaotic terrain provide evidence that water also played a role in its development. One of the most critical observations that any of these mechanisms must explain is how the rock was removed from the interior of Valles Marineris and where it went. There is no evidence for significant transport of debris within the canyons by either wind or water, nor are the obvious large-scale deposits at the ends of these canyons of materials transported out of the canyons.

Tectonism most assuredly played an important role in the development of Valles Marineris. The linear nature of the canyons, the many parallel faults inside the canyons, and the orientation to the Tharsis rise are consistent with a structural origin. In its location on the side of Tharsis, the crust under Valles Marineris would have been stretched and pulled apart as the plateau was uplifted, causing it to develop partly as a "keystone collapse." Though less likely, it has been suggested that Valles Marineris is a product of an early failed episode of plate tectonics caused by global-scale rifting that tore apart the crust in this area. Rifting on a scale needed to produce Valles Marineris requires considerable lateral movement along huge faults at either end of the canyon system. Large faults, similar to the San Andreas fault, would be easily recognized but have not been found. It has also been suggested that the canyons initially may have formed by collapse due to withdrawal of water in the area, much as the chaotic terrain formed. Though this mechanism may explain where some of the materials went that were originally inside the canyons, it requires an enormous amount of water and a mechanism to get rid of the rest of the materials.

It has also been suggested that the origin of Valles Marineris may be tied to the extensive volcanism in the Tharsis region. Some scientists have suggested that the volcanoes of Tharsis were fed by magma generated beneath the canyon. In this hypothesis, the magma was transported through a system of subsurface conduits to a location under Tharsis, where it was erupted. Withdrawal of the magma from under the Valles Marineris region subsequently resulted in regional subsidence. This hypothesis would explain what happened to the missing materials from inside Valles Marineris and where the magma came from to build Tharsis. But it requires an unreasonably large and complex conduit system to transport a huge amount of magma the long distance to Tharsis without substantial volcanic eruption along the way.

Elysium

Elysium is the second largest volcano/tectonic complex on Mars. By any measure it is enormous. It is a broad rise, about 2,000 km (1,200 miles) across and 5 km (3 miles) high. Though much smaller than Tharsis, it shows many of the same volcanic and tectonic characteristics. Like Tharsis, Elysium is thought to sit over a large mantle plume that supplies it with heat. Elysium has been active for a considerable part of the planet's history and shows evidence of having developed earlier than Tharsis.

Elysium is also the home of three huge, central-vent volcanoes. These are enormous stratovolcanos that sit on the Elysium rise and are surrounded by extensive lava plains. The largest of these stratovolcanoes, Elysium Mons, is about 500 by 700 km (300 by 420 miles) and 13 km (7.8 miles) high (Figure 42). In contrast to the shield volcanoes that are composed of stacks of lava flows, Elysium Mons appears to be composed of both ash deposits and fluid lava flows. Local concentric faults have formed on its flanks, produced by loading of the lithosphere under its enormous weight.

Hecates Tholus, a dome-shaped volcano 170 km (102 miles) in diameter and 6 km (4.2 miles) high, is located north of Elysium Mons. Its flanks are highly dissected by numerous small channels. These small channels were probably cut into easily eroded surface materials, such as ash, by water-lubricated mass flows also composed of ash. There is a near absence of impact craters around its summit. This has been interpreted to indicate that eruptions have occurred recently in this area.

Albor Tholus, the smallest of the large volcanoes on Elysium, is located south of Elysium Mons. It is about 150 km (90 miles) across

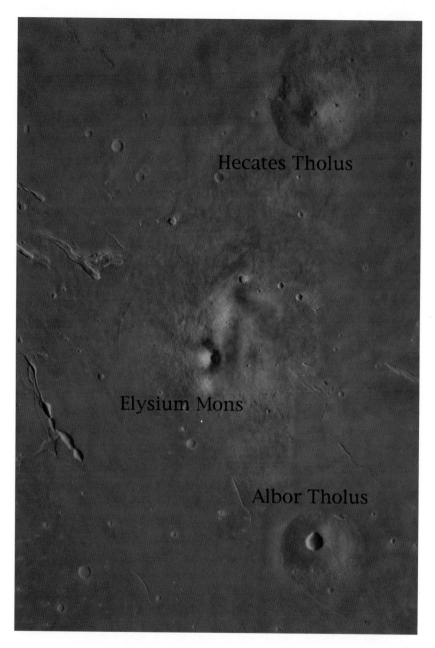

Hecates Tholus

Elysium Mons

Albor Tholus

Figure 42. The Elysium region contains the second largest volcano/
tectonic complex on Mars. Elysium Mons is the large mountain in the
center. It is about 500 km (300 miles) across. Hecates Tholus is the dome-
shaped volcano to the north and Albor Tholus is the smaller volcano to the
south. Elysium Fosse, shown on the left, are volcano/tectonic troughs and
parallel faults, some of which connect with channels to the west. Faulting
in the Elysium Fosse may have released subsurface water that carved
these channels. (Courtesy NASA/U.S. Geological Survey) (PIA00412)

with a large summit caldera about 30 km (18 miles) in diameter. The relative size of its caldera suggests that it may be partially buried by younger volcanic materials in the area.

There are also several troughs in the northwestern part of Elysium. These troughs start out as grabenlike features that are transformed into more sinuous troughs, showing evidence of disgorging water or very fluid lava. Volcanism and faulting in this area may have mobilized deposits of subsurface water or ice to form these channels.

Though Elysium is enormous, and its effects on the crust and lithosphere are locally important, they are dwarfed in comparison with the effects of Tharsis. Stresses produced by the loading of Tharsis are so large that they have an influence on the stress system in the Elysium region. Most of the faulting in the Elysium region shows these effects. For example, Cerberus Rupes shows Tharsis's influence: though it originates in the Elysium region as a southeast-trending fracture system, it changes to an easterly trend system as it nears Tharsis.

Impact Basins

Though the crustal dichotomy and Tharsis have dominated the geologic history of the crust, huge, ancient, multiring impact basins (Figure 43) have also played an important role in the history of Mars. The presence of these basins often affects regional stress patterns, exerting control on the orientation of young structures in nearby regions and opening conduits for volcanic material to reach the surface.

The most important of these great basins are Hellas, Argyre, Isidis, and the ancient buried basin Utopia. As with impact basins on other planets, the young basins include several interior nested rings of mountains and systems of radial and concentric fractures. These mountain rings and faults resulted from the accommodation of the lithosphere to the formation and presence of the basins. Rolling deposits of material ejected from these basins during their formation thin rapidly away from the outer edge of the inferred basin rims.

The crust beneath the large impact basins Hellas, Argyre, Isidis, and Utopia shows various degrees of thinning. This thinning may be due to excavation and mantle rebound resulting from the impact process. The largest basins, Hellas and Utopia, also show the same

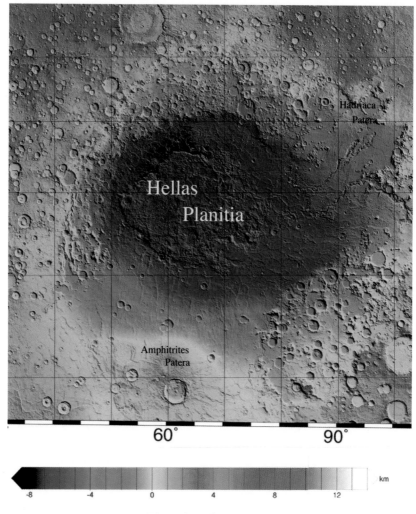

Elevation (km)

Figure 43. The Hellas Basin is an enormous impact crater. It is about
1,800 km (1,080 miles) across and 8 km (4.8 miles) deep. Several large
volcanoes are found along or near the rim of Hellas. These low volcanoes
probably formed on the giant faults that formed in concentric rings when
the basin was blasted into the side of Mars. The interior contains a
complex of plains (Hellas Planitia) deposits that may be a mixture of lake,
river, glacial, windblown, and volcanic deposits. The image is color coded
for elevation, with blue being lowest elevation and red being highest.
(Courtesy NASA/Goddard Space Flight Center/Jet Propulsion Laboratory)

subsurface structure that is characteristic of lunar mascon basins and the Chicxulub impact basin (of dinosaur extinction fame) on Earth: characterized by a thinned central region surrounded by a ring of thickened crust. This ring structure is probably due to flow of materials in the mantle caused by its rebound from the impact and excavation process.

Highlands Paterae

Not all the large Martian volcanoes are shield or stratovolcanoes or have developed in volcanic/tectonic centers. Several huge, low-relief, central-vent volcanoes, called paterae, are found in the southern highlands. Most patera have formed along the fracture systems that ring the Hellas impact basin. These volcanoes are characterized by complex central calderas, low shield shapes, radial channels, and radial ridges. They are thought to be composed mainly of ash flow deposits.

Highland paterae are the oldest of the giant volcanoes preserved on Mars. They represent a transition from early plains-forming volcanism to large discrete volcano/tectonic centers. Their old age and low relief reflect an early period in the history of Mars when volatiles inside Mars were more plentiful and as a result gas-driven explosive volcanism was more common. Alternatively, their shapes may reflect compositional differences in erupted materials produced by compositional differences in the magma source regions beneath the southern highlands and northern plains.

Three huge, ancient, low-profile paterae, Hadriaca Patera (Figure 44), Amphitrites Patera (Figure 45), and Tyrrhena Patera (Figure 46), as well as several smaller volcanoes have developed along faults that ring the Hellas Basin. Each of these large volcanoes has channels carved in their easily eroded flanks. Some appear to have been cut by water and others from the action of lava. For example, the long, continuous channels that dissect the flanks of Hadriaca Patera were probably formed by groundwater sapping or surface runoff. In contrast, a large channel that extends southwest from the caldera of Tyrrhena Patera is thought to be a lava channel that supplied lava to the numerous 100-km (60 miles) -long flows on that flank of the volcano.

Only one patera is found in the north, Tempe Patera. Though in

Figure 44. Hadriaca Patera is a low-profile volcano located on the
northeast rim of the Hellas Basin. Like the other paterae, Hadriaca is
older than the high-profile volcanoes common to Tharsis and may reflect
different geologic conditions at the time of eruption. South of Hadriaca
Patera, Dao Vallis begins as a steep-walled depression 40 km (24 miles)
wide but forms a much shallower channel that extends for 800 km
(480 miles) southwest into the Hellas Basin. The channel was very likely
carved by groundwater release from the subsurface by volcanic activity.
(Courtesy NASA/U.S. Geological Survey) (PIA00415)

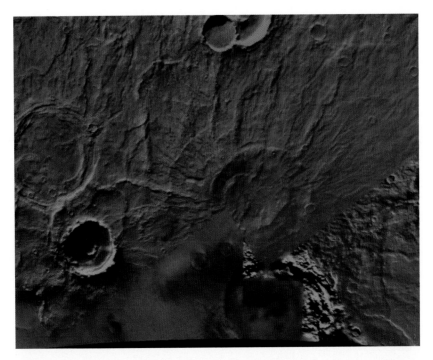

Figure 45. Amphitrites Patera is a low-profile volcano, 110 km (66 miles) in diameter, that appears to be the source of lava flows that constitute the ridged plains to the south. The characteristic ridge pattern around this volcano can be traced northward for 400 km (240 miles) into the Hellas Basin. (Courtesy NASA/U.S. Geological Survey)

the Northern Hemisphere, Tempe Patera is also located in highlands terrain. It looks very similar to Tyrrhenna Patera except it is much smaller and slightly older. The smooth deposits that surround its 16-km (9.6-mile) central depression have been extensively degraded by radial, poorly defined shallow troughs or channels.

Small Volcanoes and Lava Plains

Though not as spectacular as the enormous central-vent volcanoes, small volcanic edifices (i.e., cinder cones, lava domes, spatter cones, low shields, and dikes) are common and have also played an important role in the volcanic history of Mars (Figure 47). These features closely resemble their terrestrial counterparts and are commonly

Figure 46. Tyrrhena Patera is a broad, low-relief volcano. The scene shows a central circular caldera 12 km (7.2 miles) in diameter surrounded by a fracture ring 45 km (27 miles) in diameter. The fracture ring is surrounded by highly eroded, rugged, outward-facing cliffs, as though successive flat layers had been eroded back. Several flat-floored channels are cut into the terrain. These channels may have been carved by lava. (Courtesy NASA/U.S. Geological Survey) (PIA00421)

Figure 47. Cone-shaped hills on lava flows in southern Elysium Planitia. These craters are thought to be small volcanic cones formed by explosions due to the interaction of molten lava with a water-rich surface. The view in this scene is 3 km (1.8 miles) across. (Courtesy NASA/Jet Propulsion Laboratory/Malin Space Science Systems, Incorporated) (PIA02341)

only a few kilometers across. Clusters of these features are found in the northern plains and southern highlands, in particular in the Acidalia, Utopia, Aeolis, Thaumasia, and Tempe regions. Their small size and the effects of erosion make it difficult to identify them conclusively as volcanic features instead of impact craters.

Though volcanic edifices are common on Mars, lava plains dominate its surface area. They compose nearly 60 percent of its total area. Some lava plains are among the most ancient deposits on Mars (nearly 4 billion years old), but others are among the very youngest (younger than a few million years old).

Some lava plains are composed of single, simple flows that may be up to 1–2 km (0.6–1.2 miles) thick and extend over great areas. This type of lava plain resembles the lunar maria and flood basalts

Figure 48. Lava flows on the lower slopes of the northwest flank of Olympus Mons. As on some terrestrial lava flows, leveed channels are common on these flows. (Courtesy NASA/Jet Propulsion Laboratory/ Malin Space Science Systems, Incorporated) (PIA02080)

on Earth. Such plains are generally formed when the eruption rate is high and continuous. Other lava plains are composed of stacks of multiple overlapping lobes from numerous smaller flows. Such plains are commonly found around the periphery of the shield volcanoes (Figure 48) where sporadic, relatively low rates of eruption are common. Typically, the source vents for lava flows are fractures hidden beneath their own flow materials.

Chapter 5

The Surface of Mars

In the previous chapter, we began our exploration of Mars with its interior, its characteristics, and how they have changed with time. In this chapter, we will step out onto the surface and explore what is there and how it got that way. This step will start us on a tour of the great variety of types and ages of landforms and landscapes found on the surface of Mars. These features are a record of the rich evolutionary history of Mars.

Though collecting information about a planet's surface is easier to do than collecting information about its interior, the task is still not easy. Although the surface of a planet may be exposed for all to see, and its physical and chemical characteristics can be measured directly, their measurement generally requires transporting sophisticated scientific instruments aboard spacecraft to the planet. For Mars, this task has been difficult, with disappointing failures along the way as well as successes with huge payoffs.

Each time a new spacecraft has visited Mars and peered at the surface with a new generation of scientific instruments, our perception of Mars has radically changed. At the beginning of the space age and on the eve of mapping Mars, most scientists agreed with Jack McCauley, of the U.S. Geological Survey and a Mariner 9 team member, when he joked at a press conference that Mars was little more than "the Moon with polar caps." We now know that this is far from truth.

Surface Processes: Mars Style

Of the planets, only Earth has a greater variety of types of landforms and landscapes than the rugged desert-scape surface of Mars. This diversity in landforms reflects the complex interplay of a great range of processes that operate on the surfaces of both planets. The relative importance of each of these processes varies with time and location on both planets.

The surface of a planet can be changed through processes that fall into one of three categories: exogenic processes, in which the action of the wind, water, ice, gravity, and solar radiation erode, weather, and remove rock; endogenic processes that deform rocks due to heat from the interior (covered in chapter 4); and external processes that originate from outside Mars, such as impact cratering. Of these, exogenic processes have most dramatically affected Earth's surface, strongly driven by abundant solar energy and liquid water. Because of its dynamic surface environment, Earth is unique among the planets; its landforms are built and destroyed at a rate so high that any landform older than several hundred thousand years is considered old. In contrast, on most terrestrial planets where water is absent, mass wasting (a process of transport, such as landslides, in which gravity pulls material downslope) and meteoroid bombardment are the most important erosion agents. As a result, on these planets, the rate of weathering and erosion is dramatically lower and many landforms remain essentially unaltered for billions of years.

Although the surface of Mars contains many landforms that are billions of years old, there are others that have been eroded or buried much later in the history of the planet, recording evidence of a dynamic surface environment. With liquid water currently unable to exist at the surface of Mars without freezing and evaporating, both the physical and chemical breakdown of rock at the surface occurs comparatively slowly. Under current conditions, chemical breakdown (chemical weathering) of rock on the surface of Mars is mainly through reaction with water vapor in the atmosphere, water from volcanic activity, or impacts that tap into subsurface water-rich rock. The physical breakdown of rock (physical weathering) on Mars occurs generally when surface materials are ground up during landslides, faulting, impact cratering, or volcanism. On Earth weathered materials are transported primarily by the wind, water, ice, or mass wasting. On Mars the relative importance of each of these mechanisms is still a matter of debate, though it is clear that the wind and mass wasting are currently the dominant exogenic processes. Remarkably, in a few locations there is evidence that currently water may intermittently bubble to the surface to contribute to the weathering and erosion of the Martian landscape.

For the rest of this chapter, we will discuss Martian landforms and landscapes, grouped by the processes that form them. In particular, we will discuss what these landforms tell us about what Mars was, what it is now, and how it got that way. In the next section, we will take a close look at the surface of Mars in the three places that our spacecraft have landed.

Up Close and Personal

"Rocks, rocks, look at those rocks," Matthew Golombek howled in excitement as the first Mars Pathfinder image unfolded on the monitor in front of him. As other Mars Pathfinder team members crowded around the monitor with him, they were staring at a scene that looked remarkably like those seen by the Viking teams over 20 years before them. The Viking 1 and 2 and Mars Pathfinder Landers provided our only close-up views of the surface of Mars. All three of these spacecraft landed on rock-strewn plains that superficially resemble one another. As a result, it would be tempting to conclude that all of Mars looks like these places. This assumption most likely is wrong. These sites were picked with landing safety as the primary concern. As we learned earlier, the only data available to support landing site safety certification, such as pictures from orbit, have provided little insight into the details of the surface and have led us to land at places that look the same. As our data sets become increasingly sophisticated, our understanding of Mars becomes deeper and we will begin to explore more complex, diverse, and different terrain.

Two of these landers, the Viking 1 and Mars Pathfinder Landers, set down in the southern Chryse Planitia, Viking 1 on the lava plains in its southwestern corner and Mars Pathfinder in a broad flood channel carved in its southeastern corner. Chryse Planitia is in the Chryse Basin, a huge ancient impact scar, partially flooded by lava plains. In this respect, the Chryse Basin is similar to lunar mare basins visited by the Apollo missions. But unlike the lunar mare basins, Chryse has also been the location of a series of enormous flash floods. These floods originated in collapsed terrain to the south in the Valles Marineris region. Water from these floods breached the basin's rim in several places, scouring enormous channels. The channels can be traced on the surface for a short distance into the basin, where they disappear under the lava plains. Remarkably, these buried channels are so large that their gravity signature can be traced for several hundred kilometers farther into the basin from where they disappear under the lava plains.

Viking 1

The area surrounding the Viking 1 Lander is a rolling plain covered by dust and numerous rocks (Figure 49). In addition to rocks strewn about the surface, bedrock appears to outcrop near the lander. Several impact craters are visible from the lander. These may be the source of most of the rocks found at the site, probably excavated from the surrounding plain and scattered about by the formation of these nearby impact craters. The rocks have a variety of shapes, textures, and sizes. They range from several centimeters in diameter to up to several meters across and are similar in size distribution to the rocks measured in the high-resolution Lunar Orbiter pictures and found around Tycho, the Moon's youngest large crater. Remarkably, no rocks smaller than several centimeters across are visible at this site, perhaps due to wind abrasion that selectively destroys the smallest rocks.

Compared with the Apollo landing sites on the Moon, why are there so many rocks lying on the surface at this site? The answer may lie in the relative numbers of small impact craters. On the Moon, rocks and boulders are produced by the formation of large craters and then smashed and broken up by the constant rain of small meteoroids. Unlike on the Moon, few small craters are found at any of the Mars landing sites, though there are a number of subdued depressions a few meters across within view of the Viking 1 Lander that may be eroded small impact craters. This paucity of

Figure 49. The Viking 1 landing site, located in Chryse Planitia, is covered by numerous rocks and drifts of fine-grained material. (Courtesy NASA) (PIA00393)

small craters on Mars is probably due to the same reason small craters are not seen on Earth—meteoroids that produce these craters are generally filtered out by the atmosphere. As a result, on Mars relatively few small meteoroids reach the surface to break up the rocks produced by the larger impacts.

The composition of the rocks at the Viking 1 site is unknown, though some inference can be made from the data that exist. Some of the rocks found on the surface are pitted and resemble those commonly found in the upper zones of terrestrial lava flows of basaltic composition (Figure 50). In these terrestrial flows, such pits are the result of the preservation of gas bubbles contained in the lava as it solidifies. By analogy, if the holes in the rocks at the Viking 1 site are, indeed, such gas bubbles, then these rocks have a basaltic composition. This is also consistent with their dark gray color—the same color as basalt. In addition, the rocks at the Viking 1 site are clearly derived from the surrounding dark gray lava plains and these plains appear similar to lava plains on Earth and the Moon composed of basaltic rock. Though there is also some evidence that andesitic rock is common in the Northern Hemisphere of Mars, there is other evidence that the composition of some of these rocks is closer to basalt than to andesite (see chapter 4 for details). Considering all the data collected, at this point our best guess at the composition of the rocks at the Viking 1 site places them as being basalt.

Figure 50. Shown here in the first picture ever taken on the surface of Mars, pitted rocks are common at the Viking 1 landing site. The pits are thought to be vesicles, gas bubbles preserved in the rock when it solidified. The large rock in the center is about 10 cm (4 inches) across. At right is one of the lander's footpads. (Courtesy NASA) (PIA00381)

Figure 51. View of the Mars Pathfinder landing site in Ares Vallis taken from Mars Pathfinder. At left *(left view)* is one of the lander's petals and low-gain antenna mast. On the right *(left view)* is the fully deployed ramp Sojourner used to descend to the surface. At top left *(right view)* Sojourner

In contrast to the dark gray rocks, the dust at the Viking 1 landing site is reddish and bright. This dust is probably composed of clay minerals derived from the rock. It is thought to owe its reddish color to oxidation (rusting) of the iron in these clays. The dust at this site appears to be the same everywhere and shows little variability in its characteristics. It commonly forms drifts on the downwind side of the rocks at this site, as well as stand-alone drifts and drift complexes, some as much as 10 m (33 feet) across. In spite of being composed of dust, these deposits show remarkable stability to wind erosion. None of these large drifts changed appreciably during the 6 years of observation by Viking, although smaller dust deposits showed substantial changes in their shape after each of the three major dust storms that passed over the site.

Mars Pathfinder

Upon landing, Mars Pathfinder peered out onto a rugged, rock-strewn plain at the mouth of Ares Vallis in southwest Chryse Planitia (Figure 51). The effects of the floods that carved Ares Vallis are seen everywhere at the site. Numerous low-relief parallel ridges and valleys cross the landing site, giving it a rolling appearance. These features are aligned in the direction of Ares Vallis and are probably scours or streamlined sandbars produced during the floods that formed the channel. As with rocks found in terrestrial streambeds,

appears to the left of the rock. On the horizon the double "twin peaks" are visible about 1–2 km (about a mile) away. The terrain surrounding Mars Pathfinder was carved into long, low ridges by the floods that formed Ares Vallis. (Courtesy NASA) (PIA01466)

some rocks at this site appear to be aligned in the direction of the flow down Ares Vallis. These rocks may have originated elsewhere (upstream) and were washed to their current location by the flood.

Not all rocks are aligned. Some rocks are randomly distributed around this site and have about the same size frequency distribution as rocks at the other Mars landing sites. These rocks are most likely derived from the nearby impact craters and are presumably the same composition as the plains.

As at the Viking 1 site, some rocks are pitted, though others are massive and some even appear to be layered. All are probably volcanic rock derived from the surrounding plains. Although this is generally regarded as the simplest explanation, some scientists question this interpretation and argue that the pitted and layered rocks may be sedimentary rocks. These scientists contend that the layered rocks might be sandstone or siltstone deposited in running water. Even more controversial, they argue that some of the pitted rocks are sedimentary rocks called conglomerates (Figure 52). This type of rock forms when flowing water deposits pebbles or cobbles in a matrix of sand, silt, and clay that later is compressed and cemented into rock. There is speculation that the pits in these rocks are voids left when embedded pebbles weathered out.

Mars Pathfinder carried a more limited scientific payload than did Viking, but its instruments were much better suited for making compositional measurements. Using its alpha-proton X-ray spec-

Figure 52. Pitted rocks are common at the Mars Pathfinder landing site. The origin of the pits is controversial and has tremendous implications for the environment where the rocks formed. They could be either gas bubbles preserved in volcanic rocks as they solidified at the tops of lava flows or holes left by pebbles as they weathered out of a type of sedimentary rocks called conglomerates. (Courtesy NASA) (PIA01570)

trometer to analyze nine rocks at its landing site, it found them all to be composed of silicon-rich volcanic rock, similar to basaltic andesites (see chapter 4). This type of rock is common in Earth's continents, produced by considerable volcanic processing of a planet's crustal materials.

Though these alpha-proton X-ray spectrometer analyses identified only one type of rock, spectral analysis of Mars Pathfinder pictures indicate that there are at least two types of rock at the site. Where did these two types of rocks come from? Considering the thermal imaging spectrometer measurements from Mars Global Surveyor, the basaltic andesites are derived from the plains around Mars Pathfinder, tossed to their current locations by impact crater events. The basalt is thought to be derived from the highlands, washed to the Mars Pathfinder site by the great floods that carved Ares Vallis.

Remarkably, spectral analysis also suggests that there are at least four soil types at the site, as well as duricrust, lightly cemented fine-grained sediment (Figure 53). These soils appear to be end members of two major types, bright soil and dark soil. The bright soil is most likely extensively oxidized and weathered clays, and the dark soil is thought to be composed of tiny less oxidized and weathered rock fragments. Some of the soil particles were picked up by magnets on the lander. These particles appear to be coated with magnetic minerals. The magnetic mineral coatings are thought to have formed by the precipitation of iron from water containing iron minerals.

Figure 53. Different types of soil at the Mars Pathfinder landing site shown here as different colors. In this picture Sojourner Rover straddles a dune containing dark, sand-size granules. The tracks of Sojourner also uncovered bright reddish dust and dark reddish soils. (Courtesy NASA) (PIA01132)

As at the Viking 1 landing site, evidence of wind activity is everywhere at the Mars Pathfinder site. For example, poorly developed sand dunes are found. These indicate that the wind has been efficient enough to move sand-size particles at this site. Numerous wind tails also are common adjacent to rocks and pebbles, produced by stripping and deflation of fine-grained soil. The orientation of these dunes and wind tails is similar to the much-larger-scale bright wind streaks seen in Viking orbiter images. This suggests that all these features formed in the current wind regime.

Does this mean that the current wind regime is a permanent feature of Martian atmospheric circulation? The rocks say no. Some rocks show fluting of their surfaces, marks of erosion by the wind. These fluted rocks, called ventifacts, provide an indication of wind direction. The orientation of the flutes at the Pathfinder site is about 80° different from the direction indicated by the bright wind streaks, wind tails, and dunes. The discrepancy between the flutes and these other wind-produced features is probably the result of a change in wind regime between their formation.

Two small mountain peaks, dubbed "twin peaks" by the Mars Pathfinder science team, can be clearly seen from the lander (see Figure 51) and in orbital pictures. These peaks are located about a kilometer from the lander. They are eroded remnants of the plains that somehow survived the floods. There are several horizontal bands of light-colored material near their summits. Because of their distance from the lander, it is impossible to resolve whether these are layers of bedrock or lines of boulders left by the floods that carved the channels.

Viking 2

Though Viking 2 also touched down in the northern plains of Mars, it set down in a different geographic region and geologic setting than the other two landers—the sparsely cratered plains of Utopia Planitia. In spite of this, its rock-strewn surface superficially resembles the other sites (Figure 54). These rocks were most likely derived from the surrounding plains, dug up by the impact craters in the area.

The Viking 2 landing site is relatively flat and no bedrock outcrops or large drifts of dust are evident. The most significant topography is several flat-topped bluffs on the horizon east of the lander. They may be ejecta deposits from Mie, an impact crater 90 km (54 miles)

Figure 54. View of the Viking 2 landing site in Utopia Planitia. Large rocks litter the site. Like the rocks found at the Mars Pathfinder and Viking 1 sites, many of the rocks are porous and contain numerous pits. A shallow trough filled with fine-grained sediment runs from the lower center of the picture to the upper left. Several trenches dug by the Viking sample arm appear in the foreground. The cylindrical object on the surface near these trenches is the protective covering jettisoned from the sample arm after landing. (Courtesy NASA) (PIA00364)

in diameter about 170 km (102 miles) to the southwest, or some of the eroded topography of Utopia.

Unlike the other two sites, a polygonal network of small, shallow interconnecting troughs has developed at the Viking 2 site. These troughs are found within several meters of the lander and are approximately 1 m (3.3 feet) wide and 10 cm (4 inches) deep. Their size and pattern have fueled debate about their origin. Some scientists believe that they are contraction features, similar to cracks found when wet soil dries or lava flows cool. Other scientists argue that they are cracks produced by ice wedging of the type common in the high latitudes on Earth. These scientists argue that ice wedging and other processes common to cold regions are likely because the Viking 2 landing site is in the high latitudes on Mars. At these latitudes winter temperatures are typically below the frost point of both carbon dioxide and water. These scientists point to the thin patches of water frost that formed at the landing site during northern winter as an indication that such processes are possible at this site (Figure 55).

Figure 55. A thin white coat of frost covered the Viking 2 landing site in late spring. The frost is probably composed of water. (Courtesy NASA) (PIA00571)

In places at this site, a thin near-surface crust of loosely cemented dust called a duricrust is exposed (Figure 56). This duricrust is thought to be composed of dust cemented with leached material (mainly salts). These materials are probably deposited as water and carbon dioxide migrate back and forth between the regolith and the atmosphere. The duricrust is easily broken up into clods that resemble small rocks. To the frustration of Viking scientists, attempts to analyze small rock fragments by both Viking landers failed because they were able to find and retrieve only dirt clods composed of this duricrust. The absence of rocks smaller than 1 cm (0.25 inch) at the Viking 2 site is still a mystery.

Now that we have stood on the surface of Mars (in a virtual sense) and studied the three places our machines have visited, we will extend our exploration of Mars to its entire surface by examining its

Figure 56. Fractured duricrust is exposed near the center of the picture at the Viking 2 landing site. This thin layer of lightly cemented, fine-grained sediment is about 1 or 2 cm (0.4–0.8 inch) thick and occurs at both Viking landing sites. Duricrust is thought to form by precipitation of a cementing agent due to evaporation of volatiles in the upper few centimeters of the Martian soil. (Courtesy NASA, Viking 2) (22A007)

landforms and landscapes and considering how and why they have changed through time.

The Windswept Surface

The thin, dry atmosphere of Mars supports a surprising range of chemical and dynamical processes that have had dramatic effects on the surface (see chapter 6 for more details). Early astronomers noting rapidly changing surface markings on Mars correctly concluded that these markings were clouds. These clouds could be divided into two major types, white clouds and clouds "having the color of Mars." Astronomers concluded correctly that the white clouds are condensation clouds and the reddish yellow clouds are dust storms. By carefully tracking clouds, wind speeds were estimated in the range of 10 to 50 km/hr (6 to 30 miles/hour), with peaks of over 100 km/hr (60 miles/hour) during the early phases of some of the largest dust storms. Early astronomers could do little more than speculate about the effects of the winds on the surface of Mars. Most felt that such high wind speeds would result in erosion of the surface in much the same way as occurs in deserts on Earth. Attempts were made to calculate the effects of erosion by the wind (called eolian erosion after the Greek god of the wind, Eolis) on Mars, but the experiments were seriously hampered by the lack of information about the surface environment. Eventually, missions were sent to Mars to collect these data so that we could gain a better understanding of the wind's effects on the Martian surface.

Starting with Mariner 9, pictures of the surface of Mars have shown a rich assortment of topographic features that strongly resemble those caused by the wind and found in the deserts of Earth. But considering the age of the Martian surface, the wind has affected it little. This has led scientists to speculate that something about the Martian environment slows eolian processes.

Eolian Processes

In comparison with Mars, wind-produced landforms are rare on Earth. Why is this, when Earth has a much more substantial atmosphere than Mars? The answer ultimately centers around the lack of abundant water on the surface of Mars instead of its atmosphere. Surface processes dominated by water (and ice) operate at a much higher rate than do eolian processes. On Earth there is abundant liquid water and ice, but on Mars there is little ice and essentially no liquid water on its surface. Consequently, on Mars eolian processes have little competition from other surface processes.

Added to this, the windblown sediments on Earth and Mars are different. They generally have different average grain sizes and different compositions. On Earth windblown materials are generally composed of sand-size quartz grains that chemically weather out of silicon, aluminum-rich crustal rock. Quartz is a common mineral in rocks that form Earth's crust. It is particularly tough and resistant to weathering and abrasion, and it rarely wears down to dust size. But quartz is rare on Mars. Instead, windblown materials on Mars are generally composed of dust-size grains of iron-rich clay. The grains are the weathering products of the abundant iron- and magnesium-rich volcanic rocks commonly found on the Martian surface. These iron-rich minerals are particularly susceptible to weathering and abrasion on Mars. Though these materials typically weather out of the parent rock as large particles, they quickly pass through the sand sizes to become dust particles.

Here lies another mystery. If most types of rocks found on the Martian surface quickly weather to dust, why are there so many sand dunes on the Martian surface? Whether on Mars or Earth, dunes owe their size and shape to processes that require sand-size particles. Sand-size particles on Earth are rock fragments typically composed of quartz, but there is little or no quartz-bearing rock on Mars to produce quartz sand. Remarkably, the answer may be that on Mars quartz may not be needed to make rock fragments that are sand size; instead they may be made of dust or, more correctly, dust aggregates.

At least three ways have been proposed to create sand-size grains from dust particles in the Martian environment. Experiments conducted by Ronald Greeley and his colleagues at Arizona State University and Ames Research Center using special Mars simulation wind tunnels have found that when dust particles are transported by the wind, they rub together and generate an electrostatic charge. These electrostatic charges can draw dust grains together to form small balls of dust the size of a sand grain, which suggests to Greeley that on Mars "sand-size particles could be aggregates of dust particles, bound electrostatically." Greeley has also offered an alternative way of making dust aggregates. He has suggested that "Such particles could be cemented by salts and generally related to duricrust." Although this process could produce dust aggregates, it requires that enough duricrust be continuously produced and broken up to generate the huge amount of sand needed for the myriad of dunes seen on Mars. Greeley does not favor this explanation because of the high amount of water and other volatiles required to deposit salt cements in the soils and the lack of a process to break up the duricrust rapidly into sand-size grains. In a series of experiments at Jet Propulsion Laboratory, Steve Saunders and his colleagues simulated the dusty atmosphere and surface in the Martian polar regions. They found that in the cold, dusty atmosphere above the poles, dust aggregates could be produced when frost condenses on airborne dust particles, sticking them together to produce frozen interlocking grains. As the interlocking grains accrete more grains, they grow to sand size and precipitate onto the surface. Because of the interlocking nature of the grains, these particles hold tightly together even after the frost sublimes.

Dust aggregates may hold the key to other mysteries, including how sand can be transported long distances on Mars. Dust aggregates can form in place from dust transported globally by dust storms. Adding to this, the low-density dust aggregates have about the same cross-sectional area as the higher density sand-size rock fragments, making the dust aggregates easily caught up by the wind and blown across the surface. As with quartz sand, when these aggregates are swept across the surface, it is either by rolling or bouncing along. This process of bouncing across the surface is called saltation and is the most common way for sand to be transported by the wind. For quartz sand this is no problem, but for dust aggregates the impact at the end of each bounce can be disastrous, breaking them apart. Experiments have shown that for moderate wind speeds on Mars aggregates can easily be moved by saltation and still survive the impacts without breaking apart. But if wind speeds are high enough and these aggregates are moving fast enough, the impacts will shatter them. At that point, the dust particles may be either pulled back together by their electrostatic charges into new aggregate grains or swept away into suspension in the atmosphere.

To produce dust aggregates, dust must be initially transported by the wind. Therein lies another problem. On Mars it is particularly difficult for the wind to dislodge dust from the surface. In the boundary layer of the atmosphere just above the surface, wind flow is typically smooth. Because individual dust grains are generally flake-shaped particles, they lie flat on the surface and are nearly impossible to pick up in such a wind regime. The wind simply flows over them. Wind tunnel tests show that winds must exceed the speed of sound in the Martian atmosphere to lift dust directly from the surface. Above this boundary layer, turbulence produces wind eddies that swirl down and effectively push, lift, or roll high-profile particles such as sand. Consequently, on Mars the effects of this boundary layer combine with the low density of the atmosphere to make it nearly impossible for the wind to lift dust from the surface.

Though in theory dust should remain where it is on the Martian surface, there is plenty of evidence that dust is easily lifted and moved by the wind. To explain the inconsistency between what is theory and what actually happens on Mars, several mechanisms have been proposed for how dust may be dislodged initially and lifted from the Martian surface. None has done a particularly effective job of explaining how so much dust is raised. One set of experiments conducted at Jet Propulsion Laboratory by Steve Saunders suggested that rapid changes in temperature and pressure cause the escape of volatiles from the Martian regolith, resulting in venting or fountaining that sends dust flying into the air. This mechanism requires drastic changes in surface conditions over large areas and as a result is likely to work only in the polar regions where large amounts of volatiles may be stored in the regolith.

Other scientists have speculated about the role dust devils, the small tornado-like wind storms seen in Viking, Mars Global Surveyor, and Mars Pathfinder pictures, may play in lifting dust into the atmosphere (Figure 57). Dust devils can easily raise dust particles from the surface. They produce dramatic differences in pressure between their top and bottom that act to suck dust from the surface, instead of blowing dust off the surface. But most scientists agree that there are not enough dust devils to do the job. As with many other mysteries about Mars, this one remains unsolved.

Dust aggregates may be the solution to another Martian mystery: why is the eolian erosion rate so low on Mars? Even before the exploration of Mars, theoretical studies of the erosive effects of wind-blown dust on Mars had been done. These studies were later backed up by wind tunnel tests showing that high wind velocities combined with the focusing effects of each impact caused by the tiny pointed edges of dust particles more efficiently chip away at rocks on Mars than sand does on Earth. These impacts are so efficient on Mars that old surface features on Mars should have been completely eroded away. It even prompted Carl Sagan to joke that if, indeed, these studies were true, Mars should be completely smooth and awash in eolian debris. We know that this is not the case. This is where dust aggregates enter the argument. Experimental studies have shown dust aggregates to be ineffective at sandblasting surface material. Gree-

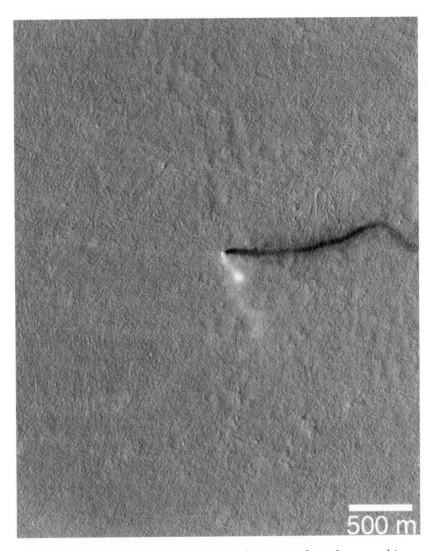

500 m

Figure 57. A Martian dust devil caught in the act. Looking down on this miniature tornado-like vortex of wind from orbit, it appears as a slightly curved and kinked, fuzzy patch with a shadow indicating its columnar shape. A thin, light-toned track has been left by the dust devil as it moved across the Amazonis Planitia region of Mars. Dust devils form when the ground heats up during the warmest parts of the day, warming the air immediately above the surface and causing it to rise. As the air rises it also spins, forming these tiny storms. (Courtesy NASA/Jet Propulsion Laboratory/Malin Space Science Systems, Incorporated) (PIA03223)

ley and his colleagues found in their abrasion experiments that "Sand-size dust aggregates, whether the dust particles are electrostatically bound or cemented together, are ineffective as abrasive agents under Martian conditions. Aggregates do not abrade rocks or rock coatings." Because the aggregates are sand-size, low-density, and low-strength grains, they either break apart upon impact or do little damage when they strike a hard surface. However, Greeley pointed out that aggregates are capable of some erosion. For example, he found that "dust aggregates can abrade their parent materials at low impact angles, suggesting that saltating dust aggregates could erode indurated soils or crusts of similar strength."

The Wind at Work

The effects of the wind are found everywhere on the Martian surface. Whether carved by erosion or built by deposition of windblown materials, there are many landforms produced by the wind that can be identified easily in pictures taken from orbit. Such landforms are a record of environmental conditions when they were formed. The nature, distribution, and history of these landforms provide insight into the surface environment of Mars.

Windblown Deposits

We begin this section on windblown deposits with one of the most common surface features on Mars: dunes. There are many types of dunes found on Mars. Each has its own unique shape, caused by differences in surface conditions during its accumulation. These different types are generally about the same size and shape as their terrestrial counterparts found in the deserts on Earth. Of the types, transverse dunes and barchan dunes are the most common on Mars. Transverse dunes are strongly asymmetrical ridges that form parallel, or transverse, to the dominant direction of the strong winds. They form in areas of near-unidirectional wind. In contrast, barchans are crescent-shaped dunes found in regions where there is not enough sand to form other types of dunes such as the transverse type. As a result, barchan dunes often form along the edges of the larger dune fields, as well as being found in small, isolated fields.

Though dunes are found all over the planet, the North Pole region contains the most spectacular dune field on Mars (Figure 58). This field is an enormous sand sea, or erg, ringing the North Pole.

This erg is nearly the size of the Sahara erg, about 680,000 km² (244,800 square miles). It is mainly composed of wave after wave of north-south trending transverse dunes spaced about 500 m (0.3 mile) apart. Commonly these dunes have superimposed oblique braids or reverse patterns, indicating modification by secondary oblique seasonal winds. Though there is evidence that these dunes may be active today (e.g., wind streak changes), their orientation and shape has not changed for the past three decades. Frost may contribute to their stability by cementing their grains together. Frost has been observed in the interdune areas during winter and was probably present when the dunes accumulated. As a consequence, the stability of these dunes may be dependent on the climate at the time they formed. If they formed during cold climatic periods when frost cement was present, then they would remain stationary until the climate turned milder. During milder times the frost cement would sublimate and the particles would be free to move.

Barchans are found along the edges of the erg as well as in small, isolated dune fields in many other parts of Mars. Most of these small fields also contain transverse or other types of dunes, such as those with star shapes, called star dunes, in addition to barchans. The dunes found in these isolated fields are about the same size as those in the erg but are generally spaced farther apart, about 1 to 2 km (0.6 to 1.2 miles) compared with the half a kilometer (0.3 mile) for those in the erg. This probably is an indicator of the abundant sand in the erg.

Wind streaks and splotches are even more common than dunes. It is difficult to look at any part of Mars without finding them. Most of these features are associated with topographic obstacles such as craters. Streaks can be either light or dark, and in a few cases both are associated with the same obstacle (Figure 59). Bright streaks are thought to be thin deposits of bright dust that accumulate in the wind shadows of topographic obstacles. These are places where wind velocities are low and dust can settle out of the atmosphere. Bright splotches (irregular-shaped wind streaks) commonly form inside craters on their downwind side where winds reverse direction, dropping some of their dust load. Dunes commonly are found within these splotches. In contrast, dark streaks form when wind speed is high and local turbulence and wind shear are at the maximum. They commonly form downwind of topographic obstacles where dust is swept away, exposing the dark surface beneath. Dark splotches are common inside craters, on the upwind side where

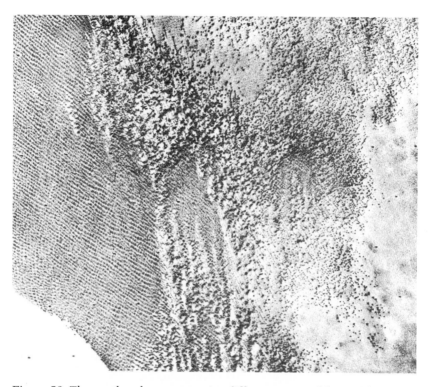

Figure 58. The north polar erg contains different types of dunes (the dark features), with transverse and crescent-shaped barchans shown here the most common. This picture shows the transition between transverse dunes and the isolated barchan dunes. The scene is 100 km (60 miles) wide. (Courtesy NASA, Viking) (58B28)

strong wind vortices sweep the surface clean of bright dust and expose the dark rock beneath.

There are other less-common types of streaks and splotches on Mars. Occasionally, a dark streak is found that originates from dark splotches. High-resolution pictures of these dark splotches generally reveal that they are filled with dark-colored dunes. The dunes are probably composed of dark particulate material, probably dark-colored volcanic ash. On the other hand, bright frost streaks have also been observed. These streaks are found in the polar regions, behind craters. Some are quite thick in places and sometimes persist long after local frost has disappeared for the season.

Streaks and splotches are thought to indicate wind direction and hence can be used to determine wind direction at the time of their

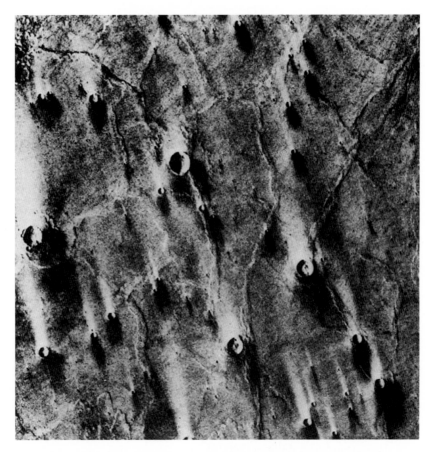

Figure 59. Light and dark wind streaks are shown here trending in opposite directions from the same topographic feature. The streaks indicate two different wind regimes at work. The dark streaks are thought to be bare rock exposed by wind scouring, and the bright steaks are deposits of dust laid down in the wind shadow of the craters. (Courtesy NASA, Viking) (553A54)

formation. They occur globally but are most common in the mid-latitudes. By comparing Mariner 9 images taken in 1972 with images taken by Viking in the late 1970s and those from Mars Global Surveyor in the early 2000s, it has been found that these features are amazingly stable. Their orientation and pattern are thought to be controlled by the high winds that occur during global dust storms. Because the global dust storms occur predominantly during southern summer, the streaks and splotches probably form only during that season. However, the pattern of dust storm activity is thought to change periodically and so will the pattern of these features. Over a much longer time scale, the orbit and axis of Mars slowly change in cycles that move the season of maximum dust storm activity. As a result, the dominant wind direction during maximum dust storm activity will shift and bring about considerable changes in patterns and direction of the streaks and splotches.

Sculpting by the Wind
Windblown particles do more than pile up in sedimentary deposits; they slowly sandblast the surface. There is evidence that throughout Martian history the surface has undergone eolian erosion at rates that vary with time and from place to place. The degree of such erosion is mainly dependent on the type of rock on the surface. If the surface is composed of solid crystalline rock, then erosion is very slow. But if the surface is composed of "soft," easily eroded rock such as volcanic ash or sedimentary rock, then the erosion rate is typically much higher.

The best example of highly eroded terrain is found in the southern part of the Amazonis region, where some of the most spectacular wind-sculpted landforms in the solar system have been incised into a massive deposit of unknown origin (Figure 60). Numerous irregular hollows, fluted escarpments, pedestal craters, and yardangs have been worn into the surface of this deposit. Yardangs are double-tapered, closely spaced sets of elongated streamlined ridges that form parallel to the direction of the wind. These landforms are considered a reliable means of determining long-term wind direction. A. Wesley Ward at the U.S. Geological Survey has studied the Martian yardangs and found that "many of the yardangs on Mars are not oriented parallel to the winds indicated by regional wind streaks." He concluded that "the yardangs were formed by different winds than those that formed the streaks." Does this indicate that the dominate wind direction changed with time?

Different types of small, wind-eroded deposits are common in other parts of Mars. Some of these appear to be erosional remnants of layered sedimentary rocks deposited in ancient lakes and eroded into streamlined landforms by wind erosion. These are discussed later in this chapter. Other deposits, particularly those in the high latitudes (especially in the south polar region), are thought to be remnants of an ancient debris mantle composed of dust (or sand) ce-

Figure 60. Eroded deposit in the Amazonis region containing well-developed yardangs. These wind-sculpted features appear in a massive, easily eroded deposit of unknown origin. They are regarded as good indicators of long-term wind direction. (Courtesy NASA, Viking) (44B37)

mented by frozen volatiles (Figure 61). These eroded deposits typically form irregular-shaped, low mesas, but remarkably none has been sculpted into yardangs. The absence of yardangs seems inconsistent with wind erosion of easily sculpted dust deposits. To explain this apparent inconsistency, scientists suggested that, during warm climatic periods, these deposits disaggregate when volatiles that cement and hold the dust grains together sublime. Because sublimation would occur over the entire surface of the deposit, the deposit is removed uniformly instead of just from its windiest surfaces. This scenario implies that the climate has changed dramatically since the time the mantle was deposited, from a time when the climate favored deposition of windblown sedimentary materials to later when

Figure 61. South polar etched-pitted deposits shown here are thought to be composed of dust cemented by ice. The etching is thought to occur as the ice sublimes and leaves the dust to be removed by the wind. This deposit appears to be layered and may be the remnant of old polar layered deposits. (Courtesy NASA, Viking) (390A90)

the climate changed and erosion dominated. More will be said about these deposits in chapter 6.

At high latitudes, particularly on the northern plains, pedestal impact craters are common (Figure 62). These craters probably have excavated a subsurface layer of erosion-resistant rock that has acted as an armor against erosion around the craters. Such craters and their blocky ejecta deposits appear to have been left standing above the surrounding terrain after the entire region was stripped and

Figure 62. Pedestal craters on the lowland plains in eastern Acidalia Planitia. These craters are thought to have been modified by the effects of deposition and deflation of surficial debris in the high latitudes of Mars. Many of the small mounds with central pits may be small pedestal craters, rather than volcanic cones, as has been proposed for similar features in other areas on Mars. (Courtesy NASA, Viking) (60A53)

eroded of its original surface. Some scientists argue that the surrounding plains are composed of volatile-rich, loosely consolidated sediment easily removed by the wind.

The slow action of the wind is only one of the processes that has shaped the Martian surface. Though Mars is currently cold and dry, there is considerable evidence that in the past running water has been at work. There is even recent evidence that water may still find its way to the surface to carve small channels and then disappear into the atmosphere. But surface processes involving both wind and water are generally dependent on the climate. As a result, these processes also provide clues to climate history and even the prospects for development of indigenous life.

Water, Water, Everywhere

By the early 1960s scientists had not only dismissed the fanciful notion that Mars was crisscrossed by Lowell's waterways, they had also nearly given up on the notion that water had existed on Mars at all. The surface environment prohibited liquid water from existing, causing it to freeze or evaporate into the atmosphere. In 1972 this view changed again. Mariner 9 pictures showed landforms on Mars that resembled dry river valleys and channels. It appeared that scientists had given up on water on Mars too soon. Harold Masursky of the U.S. Geological Survey, science leader for the Vidicon imaging experiment in 1973, stated, "The possible fluviatile channels may record episodes when water was much more abundant in the atmosphere than it is at present. Planet-wide warmer interglacial periods would release not only water locked in the polar caps but also that frozen in the subsurface as permafrost. Similar warmer and colder periods also are characteristic of terrestrial history."

For classification purposes, scientists have grouped these landforms into three major types: valleys, outflow channels, and fretted channels. Water seems to have had a role in producing all of these features. Some appear to have been carved by huge catastrophic floods that originated from massive subsurface aquifers, others by sustained but perhaps modest flows supplied by either springs or periodic rainfall. These features range in relative ages and required abundant liquid water for their development. This seems inconsistent with the evidence for cold climate conditions similar to the current climate. So it is not surprising that major controversy erupted about the origins of these landforms.

Because the Martian environment is currently cold and dry, scientists who first studied these landforms went to great lengths to explain how they could have formed under current surface conditions, without the need for water. The list of these ideas is surprisingly long and includes such processes as mass wasting, volcanism, rivers of liquid hydrocarbons, glacial scouring, and sculpting by the wind. It was even suggested that the networks of small channels were huge, intricate fracture systems resembling the small cracks common in concrete sidewalks.

In spite of how innovative these schemes may have been, and that there is still little direct evidence that it has ever rained on Mars, it appears to be inescapable that running water provides the best explanation for the production of the observed landforms. With each visit to Mars by a new spacecraft, evidence has been steadily mounting to reinforce this hypothesis. Most of the channels and valleys of Mars appear to have been cut by water fed from underground sources, requiring that these sources deliver a huge amount of water to the surface. The volume of water needed to make these landforms is much more than could be initially stored in the subsurface and suggests that the regional groundwater systems of Mars must have been recharged.

How to Make Channels and Valleys

Valleys are defined as linear troughs formed by erosion by a flowing agent, whether water, debris, or ice. In contrast, channels are defined as troughs that confine flowing materials. We find that on Earth river valleys always contain stream channels but stream channels may not have carved valleys. In addition, there is no direct relationship between the size of a stream and the size of a valley. But there is a direct relationship between the size of the stream and the size of the channel (i.e., given enough time, small streams can cut large valleys, but only large streams can cut large channels). Consequently, the size of the channel provides information about the volume of fluid that flowed down that channel.

In general, terrestrial valleys are cut in two ways: by surface runoff of water or by sapping (also called spring sapping). Surface runoff is by far the most common type of erosion by water on Earth. It erodes valleys by a combination of downcutting from surface flow and by mass wasting (sliding and slumping of the walls) of the interior valley slopes after they are undercut by the removal of materials by surface flow. These processes combine to produce valleys with diagnostic shapes, including tapered heads and cross sections that continuously increase in width downstream. In contrast, sapping occurs along fracture zones at the point where groundwater percolates to the surface to weather the rock. A valley is formed at this location because the weathered rock is more easily eroded and removed than are the surrounding materials. On Earth the surface flow of water feeds the springs from rainfall runoff. This process produces valleys with amphitheater-headed tributaries and large width-to-depth ratios. These valleys typically have straight valley segments following the fracture zones that carry the subsurface water.

For these processes to have cut the valleys found on the surface of Mars requires three conditions: liquid water, and both an adequate and long-term supply of it. To form valleys by surface runoff requires climate conditions mild enough to allow liquid water (or snow) to exist in the atmosphere and on the surface; sapping only requires that the subsurface be warm enough for liquid water to exist. The former requires mild surface conditions; the latter can occur because of either mild surface conditions or high heat flow from the interior.

Still, a huge volume of water is required to produce the observed erosion, estimated to be on the order of the equivalent of a global ocean several hundred meters deep. In considering the case of valleys that clearly owe their origin to subsurface water, the maximum possible volume of pore space in the subsurface rock is not nearly enough to hold the amount of water needed for their formation. This points toward the recharge of the groundwater system. On Earth rainfall is a critical part of the hydrologic cycle needed to recharge its groundwater system. Some scientists think that recharging the Martian groundwater system may occur by this same mechanism. They suggest that water reenters the groundwater system on Mars through infiltration of precipitation that falls during episodic climate changes producing surface conditions mild enough for rainfall or snow. They argue that the global distribution of the valleys and channels dictates this mechanism. Other scientists suggest that only the oldest valleys may require surface runoff to form and that younger valleys are eroded by water from underground sources and are recharged by processes that do not require liquid water in the atmosphere or surface.

In an effort to answer this question, Ted Maxwell and Robert Craddock of the Smithsonian Institution's National Air and Space Museum have studied the highland valleys. They have used characteristics of erosion and its effects on impact crater populations to gain insight into the origin of these small channel networks. They have concluded that "the origin of highlands drainage by rainfall versus sapping is not directly testable using crater frequency alone. Highland modification by fluvial erosion may have taken place by rainfall or by discharge from the subsurface, either of which could be responsible for crater degradation. If the drainage networks of the highlands formed by subsurface water release, then the widespread dendritic nature of the valleys requires a near-surface aquifer for their formation. However, the presence of drainage channels on small crater rims and other short wavelength features down to the limit of Viking Orbiter resolution argues for an atmospheric source rather than an interconnected aquifer that could provide a source for surface erosion on features at all scales."

Considering the current dry, cold climate that prohibits rainfall, how can the Martian global groundwater system supply enough water to produce the valley systems without recharging by infiltration of rainfall? Some scientists argue that a uniquely Martian type of hydrologic cycle may operate on Mars, even with the current climate. Steve Clifford of the Lunar and Planetary Institute and his colleagues have proposed "a mechanism whereby groundwater that was discharged on the surface during the development of valley networks and outflow channels may have been reintroduced into the subsurface to replenish the global aquifer from which it was originally withdrawn." They have suggested that any water on the surface or in the regolith near the surface will sublime or evaporate into the atmosphere and be carried to the winter polar caps where it is deposited as frost. Over time, the thickness of this frost grows to where its weight causes pressure melting at the base of the pole caps. Such melting occurs under thick glaciers on Earth and is the source of water for streams that emerge from large glaciers. Clifford argues that the liquid water beneath the caps soaks into the ground, migrating downward and outward slowly to recharge the groundwater system. This is especially important in the Southern Hemisphere, where the South Pole sits at a high elevation. Subsurface water soaking into the ground in the south polar region would trickle downslope to a lower elevation in the north, where it could break out onto the surface again. Evaporation and sublimation of this water completes the cycle. Though this process provides a mechanism that is possible, it operates very slowly and may not be efficient enough to be the entire answer to the question.

Whether this particular mechanism works or not for recharging the groundwater system, evidence has been found in high-resolution

Mars Global Surveyor pictures of recent spring-fed flow in a few locations on Mars. Michael Malin (the principal investigator for the Mars Observer camera on Mars Global Surveyor) and his colleague Ken Edgett, both at Malin Space Science Systems, Incorporated, made a remarkable discovery. They found small gullies within the south-facing walls of a small number of impact craters, south polar pits, and two of the larger Martian valleys (Figure 63). Most of these features are found in the middle and high Martian latitudes (particularly in the Southern Hemisphere). Malin and Edgett carefully examined these features and suggested that "We see features that look like gullies formed by flowing water and the deposits of soil and rocks transported by these flows. The features appear so young that they might be forming today. We think we are seeing evidence of groundwater supply, similar to an aquifer." Speculating how these features may have formed, they offer, "We've come up with a model to explain these features and why the water would flow down the gullies instead of just boiling off the surface. When water evaporates it cools the ground—that would cause the water behind the initial seepage site to freeze. This would result in pressure building up behind an 'ice dam.' Ultimately, the dam would break and send a flood down the gully."

This is an important discovery and suggests that, in spite of the current cold, dry climate, there are places on Mars where conditions are right (a source of water and temperatures high enough) for liquid water to flow on the surface long enough to produce erosion. In the next section, different types of water-produced features are discussed, such as valleys, channels, and the more controversial oceans and lakes. These features show clear evidence that water has flowed across the surface of Mars at times in its past. But whether the observed landforms were produced by surface runoff from rain or from groundwater is still a major question.

Valleys

Martian valleys are most common in the old terrains in the low to midlatitudes. These features generally lack bed forms that are direct indicators of fluid flow, although in a few instances channels have been found in the bottom of valleys. Martian valleys exhibit a wide variety of drainage patterns. The upper reaches of these valleys appear degraded and have higher drainage densities (i.e., more

Figure 63. Small gullies carved on the interior rim of a crater in the central Noachis Terra region of Mars. The slopes show numerous gullies with contributary patterns that argue for a fluid behavior during their creation. The gullies occur well down the slope where a distinctive boulder-rich layer is found. Gullies appear to start at specific layers. The sequence of layers clearly alternates between strata that either contain or erode to form boulders and layers that do not have boulders. The scene is about 6 km (3.6 miles) across. (Courtesy NASA/Jet Propulsion Laboratory/Malin Space Science Systems, Incorporated) (PIA03205)

branches) than the lower reaches. Their cross-valley profiles range from V-shaped to U-shaped valleys with, respectively, steep, nearly vertical walls and broad, flat floors. These features appear to have a great range in age.

Valley Networks
Small valleys are common in the southern ancient cratered terrain of Mars. These small valleys often connect to form networks similar to valley systems on Earth (Figure 64). Rarely do any of these valley

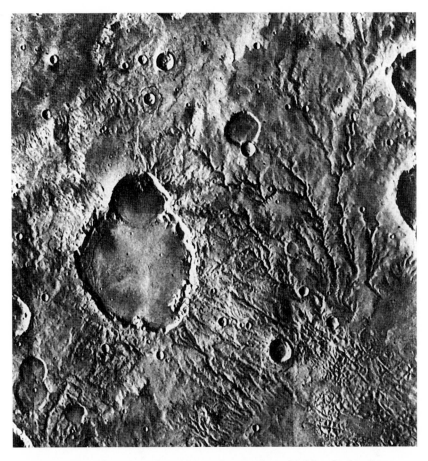

Figure 64. Heavily cratered terrain in the southern highlands that has been dissected by a network of small valleys. The smallest individual valleys are about 1 km (0.6 mile) across. The scene is about 200 km (120 miles) across. (Courtesy NASA) (P-18115)

networks exceed 300 km (180 miles) in length. Because they are found only in the ancient terrain they are thought to also be ancient. They are thought to have been carved by surface runoff or sapping or by a combination of both and as a result are generally cited as evidence for early periods of warm, wet climate when liquid water could have existed on the surface.

Typically, the individual valleys in the valley networks are a few tens of kilometers long and less than a kilometer wide (Figure 65). They tend to start small and increase in size downstream. Small fea-

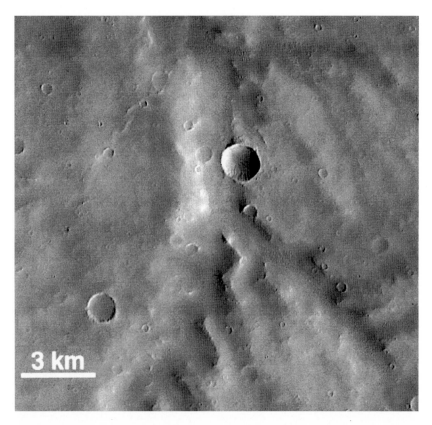

Figure 65. A high-resolution picture of a small valley carved in the Martian highlands east of the impact basin Schiaparelli. This valley is about 1 km (0.6 mile) wide. Smaller valleys are not found in this region, suggesting that groundwater processes have played a more important role in the formation of the valley systems than rainfall. (Courtesy NASA/ Jet Propulsion Laboratory/Malin Space Science Systems, Incorporated) (PIA01508)

tures such as scours and streamlined islands left by fluid flow are uncommon inside these valleys. Most appear to have a steep, nearly V-shaped cross section in upstream areas, diagnostic of surface flow. In contrast, in their lower reaches, U-shaped cross sections are common. Cross sections of this shape are diagnostic of modification by processes such as mass wasting, sapping, or glaciation. These differences are consistent with the initial formation of valleys by surface runoff and later enlargement by sapping.

This difference in formational process is also reflected in the characteristics of the valley networks. Most valley networks have relatively few branches in their lower reaches, consistent with sapping, but in their upper reaches there are a few intricately branched valleys that have tributary patterns that could be interpreted as the result of surface runoff. At the time of formation of some valleys, it is clear that surface conditions were such that modest-sized streams could flow for at least several hundred kilometers. These streams appear to have originated from and were cut by water from sources upstream, either from groundwater or surface runoff. Sapping has played only a secondary role in the formation of most valleys and may have been important in the development of only smaller tributary valleys. The discontinuous nature of some of the channels also suggests large-scale subsurface erosion and solution. The shallow, flat floors of some of these valleys do not appear to be flood plains but rather the result of later filling by eolian material or material derived from mass wasting.

With the abundant evidence for the flow of water across the surface of Mars in many regions, it is especially striking that there is no evidence of regional dissection by valleys less than a few hundred meters across. Michael Carr of the U.S. Geological Survey, the leader of the Viking Orbiter Camera Team, suggested three possible reasons for this absence of small valleys. He has speculated that "(1) erosion has destroyed all evidence of the fine-scale runoff channels, (2) no precipitation has occurred since the terrain we observe formed, or (3) infiltration characteristics and precipitation pattern were such that there was little surface runoff; all the precipitation infiltrated the ground and fed the groundwater system." He continued by saying, "The simplest explanation of the general absence of small valleys is that in the time period since the terrain stabilized, continuous, area-filling hierarchies of differently sized streams did not form, or a least were not active long enough to cause significant erosion."

There are two possible causes for small valleys not to form. The first of these is that there has been no precipitation since the current surface stabilized. The second is that there has been rainfall, but the permeability of the ground and the rates were such that most of the rainfall infiltrated into the ground instead of running off. Carr pointed out that "In the first case, all the observed valleys were cut by groundwater, and the groundwater system has not been recharged from above since the middle of the Noachian [the oldest period in

Martian geologic history dating from about 3.5 billion years ago and older], possibly because of climatic conditions that prevent precipitation. In the second case, the observed valleys were also cut mostly by groundwater but the groundwater system was periodically recharged by precipitation and infiltration. In this case climatic conditions, for at least some fraction of the period after middle to late Noachian, were such that significant precipitation could occur." But this means that either the current landscape formed after the climate turned cold on Mars and the observed valleys formed entirely from groundwater inherited from an earlier warm era before the current landscape stabilized (this is difficult to reconcile with the amount of water needed to cut the valleys), or, alternatively, warm conditions and precipitation occurred after the current landscape formed and the absence of microdissection is due to infiltration of most of the precipitation.

On some Martian volcanoes, valleys have formed that strongly resemble those of the Hawaiian volcanoes and are believed to have manifested through the same processes. Scientists have discovered that the Hawaiian valleys formed initially from surface runoff and later grew and enlarged by groundwater sapping in their lower reaches. The Hawaiian valleys are carved into porous materials, such as ash or fine-grained sediments. The Martian valleys are young, though they appear to have been formed by the same mechanism as the valley networks in the ancient heavily cratered terrain. This implies either that climate conditions permitted rainfall runoff on Mars late in its history, that springs fed by underground sources have been particularly active, or that some other mechanism not well understood was involved.

Martian valley networks form a range of patterns that resemble valley network patterns found on Earth. As in terrestrial valley networks, the types found on Mars appear to be controlled by the characteristics of the slopes. Some are carved into the rims of large ancient craters and form sets of parallel valleys. These have little or no tributary development. Other networks form rectangular drainage patterns that appear to be controlled by regional faults; others resemble terrestrial dendritic (branching like tree limbs) river valleys and have the highest drainage densities.

Another type of small channel is occasionally found on the flanks of younger Martian volcanoes. These channels appear to be lava channels or volcanic density-flow channels like those commonly associated with terrestrial volcanoes. These valleys do not form valley networks.

Trunk Valleys
In contrast to the relatively short branching valleys of the valley networks, long sinuous valleys, called trunk valleys (sometimes also called longitudinal valleys), are found meandering across the surface of the old terrain (Figure 66). These valleys vary considerably in their state of preservation from appearing fresh to barely discernible. They generally have one dominant trough, a few short amphitheater-headed tributaries in their upper reaches, and no trace of drainage features between major branches and tributaries. The short amphitheater-headed tributaries are regarded as the best evidence for sapping on Mars. The dominant trough of these valleys is generally a few tens of kilometers wide and several hundred kilometers long.

Channels are visible within some of the fresh-appearing trunk valleys. These channels may be the waterways that carried the water that cut the valley (Figure 67). As is the case with rivers on Earth, it is likely that these channels are of comparable dimensions to the stream that cut them. The shape and sinuosity of these valleys suggests that the channel that cut them probably originated at or near the heads of the current valleys. Commonly, the talus or loose rock debris derived from the slopes of opposing walls meet at the center of these larger valleys where the inner channels are located. Consequently, the width of these valleys appears to be controlled by the depth of incision and by the angle of repose of the talus, the maximum slope angle that can be supported by such loose debris. In some trunk valleys the talus meets in the middle or dunes cover the floors, presumably covering the channels.

Most of these valleys are generally quite sinuous and have a meandering pattern, though some have straight segments. The straight segments parallel regional fault systems and probably formed along faults where the ground-up rock is weak and easily eroded. In contrast to these straight segments, in some of these valleys the sinuosity is so extreme that cut-off meanders, called oxbows, have formed. These features are seen in terrestrial streams and as in the rest of the meandering stream valleys require sustained surface flow of water and relatively low gradients. The sinuosity of the trunk valleys on Mars may be secondary phenomena, inherited from the incision of

the surface by ancient streams and later enlarged by sapping and mass wasting.

Though the original sources of water for these valleys could be surface runoff, many scientists prefer a groundwater source as the fluid that cut these channels. They point to the presence of the short amphitheater-headed tributaries and other nearby sapping features in the region as evidence of abundant subsurface water. Adding to this, several of these valleys start at irregular depressions and extend as linear discontinuous valley sections. These valleys appear to have formed from a combination of surface and subsurface flow, and tectonic processes. Michael Carr has studied these valleys and concluded "The pattern is similar to that in terrestrial karst regions where solution of limestone enables large-scale, subsurface flow. Removal of subsurface material to allow collapse and formation of the depression could be the result of solution, subsurface erosion, or faulting." He added, "faulting alone seems unlikely in view of the emergence of a valley from the end of the crater chain."

Most likely, the short, stubby tributaries that branch off the trunk valleys had to compete for water with the main channel. This may be the reason for the low number and limited length of these tributaries. Once the main channel became deeper than the tributaries, local groundwater would have seeped readily into the valley to join the surface flow. Loss of this water from the groundwater system would have pulled down the local water table and deprived the tributaries of their water source.

Channels

In addition to the valleys and valley network systems, huge river channels are also found on Mars. Most of these channels are huge, spectacular landforms that are not cut valleys. There are two major types of large Martian channels, outflow and confined. All of these channels appear to have formed by periodic outbursts of water from the subsurface. Their huge size and shape suggests that they were

Figure 66. Ma'adim Vallis is a trunk valley that meanders across the Martian highlands for about 600 km (360 miles). It empties into Gustev, a crater 160 km (96 miles) in diameter. Ma'adim Vallis has many of the morphologic characteristics of terrestrial river valleys. (Courtesy NASA/U.S. Geological Survey) (PIA00414)

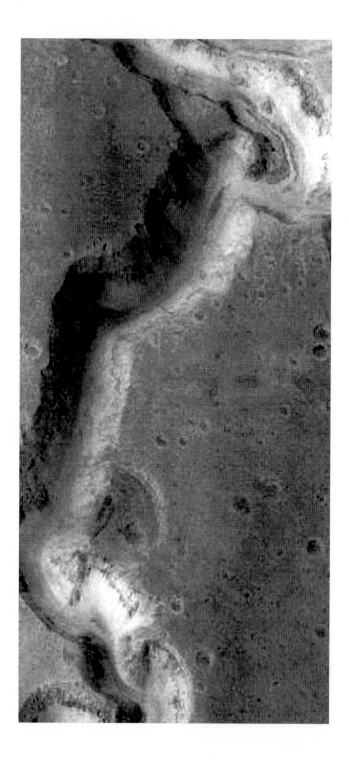

cut as a result of enormous short-lived floods. Although such enormous floods are found on Earth, they are rare.

Confined Channels

Confined channels, as they are called, are restricted to distinct, well-defined troughs. In their upper reaches, they typically start as straight troughs or in chaotic terrain, suggesting that they formed by faulting and/or collapse and withdrawal of subsurface water. In their lower reaches, these channels commonly branch into several sinuous, steep-walled channels that rejoin farther downstream (anastomosis). Most of these channels are only a few kilometers wide but can be as much as 1,000 km (600 miles) long. Though flow features generally are not found, teardrop-shaped islands are observed in the widest sections of some. These channels are best developed in the Elysium (Figure 68) and Hellas (Figure 69) regions.

Though the overall shape of these channels is more suggestive of erosion by water, they look similar to lunar rilles—channels cut by fluid lava erupted at high eruption rates. Because of the association of these confined channels with volcanic centers, volcanism probably has played some role in their formation. Volcanic heat may have been responsible for melting local subsurface ice deposits to free water and cut the channels. Alternatively, lava erosion cannot be ruled out as playing an important role in their formation.

A clue to their nature is found at the mouths of the Elysium channels. Vast (nearly 1 million km^2 [216,000 square miles]) lumpy or hummocky deposits are found at the ends of the channels. These deposits are too extensive to be only sediments eroded from the interior of the channels. They may be water-laden volcanic debris flows (called lahars) that had washed down the channels. On Earth this type of flow is common in volcanic terrain where there is abundant volcanic ash and water.

Figure 67. Nanedi Vallis, shown here, has a small channel that meanders across its floor. The small channel is about 200 m (660 feet) wide and is partially covered by dunes and landslide debris in other parts of the valley. The presence of the channel suggests that the valley might have been eroded by water that flowed through the system for an extended period of time. The scene is 5 km (3 miles) across. (Courtesy NASA/Jet Propulsion Laboratory/Malin Space Science Systems, Incorporated) (PIA02094)

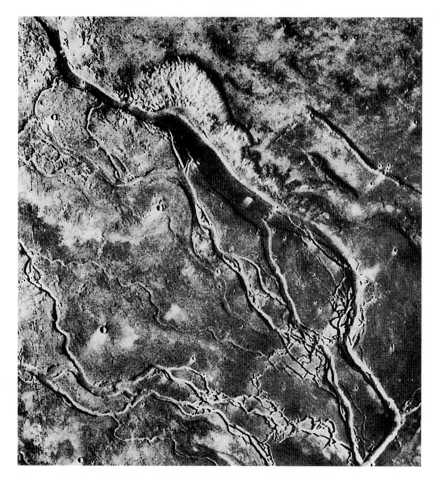

Figure 68. Confined channels west of Elysium Mons. The flow down these channels appears to have been confined to the channels only, leaving the surrounding terrain uneroded. The channels originate at grabenlike troughs to the west. (Courtesy NASA, Viking) (541A20)

Outflow Channels

Outflow channels are some of the most spectacular features on the surface of Mars. They are gigantic, channel-like landforms that can be as much as 2,000 km (1,200 miles) long and 200 km (120 miles) wide (Figure 70). They are believed to be the result of titanic catastrophic floods. As remarkable as their size and origin, these channels have developed over several hundred million years (or longer) in multiple episodes. All are younger than the valley networks.

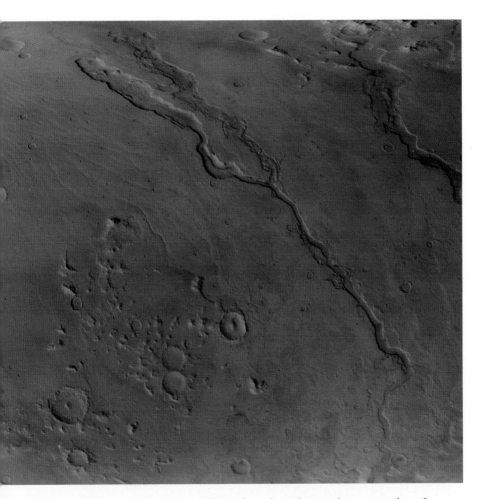

Figure 69. Three confined channels are found on the northeastern edge of the Hellas Basin. From left to right, they are Dao Vallis, Niger Vallis (joining just above the center of this picture), and Harmakhis Vallis. These valleys are believed to have formed, at least in part, by large outbursts of liquid water released from the subsurface as a result of nearby volcanic activity. These valleys are all roughly 1 km (0.6 mile) deep and vary in width from about 8 km (4.8 miles) to about 40 km (25 miles). (Courtesy NASA/Jet Propulsion Laboratory/Malin Space Science Systems, Incorporated) (PAO02810)

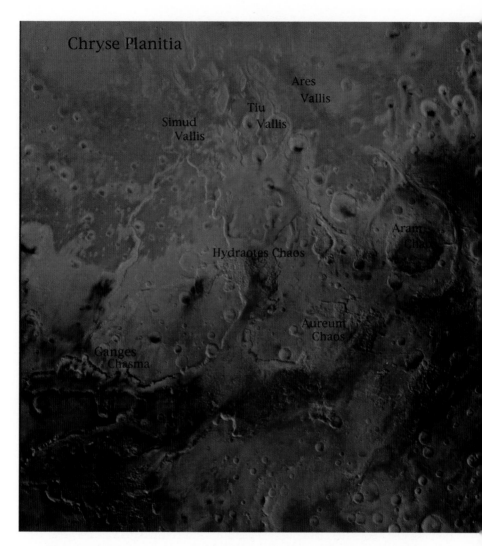

Figure 70. Regional view of the outflow channels to the east of Valles Marineris. These channels empty into the southern Chryse Basin. The chaotic terrain shown in the southern part of the picture appears to be the source of Simud, Tiu, and Ares Valles *(left to right)*. The channels are an average of 1 km (0.6 mile) deep, but can be up to 3 km (1.8 miles) deep in places. These channels are so large that their gravity signatures can be traced several hundred kilometers into the basin, long after any trace of them can be seen in photographs. The mouth of Ares Vallis is the site of the Mars Pathfinder landing. (Courtesy NASA) (PIA00418)

Evidence of flow down these channels is plentiful. They commonly contain such interior landforms as teardrop-shaped islands, longitudinal grooves, curvilinear banks, anastomosis, scouring around obstacles, and inner channels with recessional headcuts that are indicative of fluid erosion. The pattern of flow down these enormous channels is recorded as flow lines that diverge around obstacles such as craters and ridges, and become concentrated at gaps. Of all the features on Mars, outflow channels show the most unambiguous evidence of surface flow.

When outflow channels were first discovered, scientists were struck by the many similarities between them and the channeled scablands of Washington State. The channeled scablands were formed by catastrophic flooding caused by the sudden release of water from prehistoric Lake Missoula. Catastrophic flooding is extremely effective at eroding the surface. Considering this and the amount of water it would take to fill the Martian channels, each channel could have been carved in only a matter of several weeks to several months.

Most of these channels originate north of Valles Marineris within steep-sided box canyons and drain northward into the Chryse Basin where Viking 1 and Mars Pathfinder landed. They begin full sized within these canyons. They are made of terrain that is composed of a jumble of kilometer-sized blocks, called chaotic terrain (Figure 71). The jumbled nature of the blocks suggests that the chaotic terrain formed by collapse resulting from withdrawal of support beneath the terrain. The emergence of these channels full sized from the box canyons and the morphology of the chaotic terrain suggest that the terrain resulted from erosion caused by the catastrophic release of water from a huge subsurface aquifer.

Carving such enormous features in this way requires an aquifer to have contained an enormous volume of interconnected pore space. On Earth rocks that could produce an aquifer with these extreme characteristics are rare, but on rocky planets they are the foundation of most terrains. Processes such as impact cratering and tectonism produce extensive fracturing in the crusts of planets and leave their upper few kilometers ground up into a highly fractured zone, called the megaregolith. Unlike Earth (which has recycled its crust), Mars has retained its megaregolith. The Martian crust has remained stable for over 4 billion years and has recorded the geologic events that have left the crust highly fractured. As in other planets, this

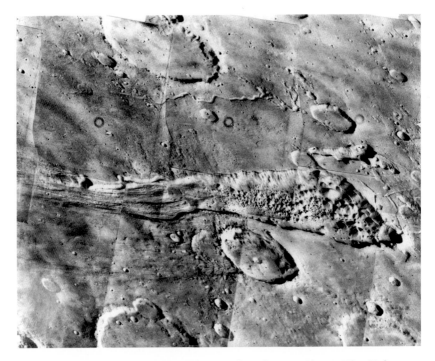

Figure 71. The chaos and channel west of Hydroates Chaos. The 20-km (12-mile) -wide channel that emerges from the chaos starts full size, suggesting a high volume of outflow over a short time. The channel connects eastward with Simud Vallis. (Courtesy NASA, Viking) (P-16983)

zone on Mars contains an enormous amount of interconnected pore space. The space forms an ideal aquifer to hold and quickly release an enormous volume of water.

A huge volume of water is required to produce these catastrophic floods. Water must be instantaneously released, implying that the water in the aquifer was under tremendous pressure. Considering the abundant evidence that subsurface water was plentiful all over Mars, why would the hydrostatic pressure be so great in the Valles Marineris region and not in others? The answer is probably connected to the building of the huge topographic high to the west, Tharsis. As the Tharsis Plateau grew, subsurface water on its flanks would have percolated downslope, driven by gravity. This would have produced a substantial hydrostatic head in the low places around the edges, such as in the Valles Marineris region. Scientists have calculated that the hydrostatic pressure in an aquifer on the flanks of Tharsis would have built up during the plateau's uplift to exceed the strength of crustal rocks in the Valles Marineris area. Such high pressures could have caused spontaneous outbursts from the aquifer, though movement along faults or formation of an impact crater probably would have breached it first.

Once an aquifer was breached, its water would have drained in an outburst that typically lasted for only about a month before it ceased. As water escaped from the aquifer, the hydrostatic pressure would fall and the rate of escape drop. Eventually, the flow rate would drop to where the escaping water would freeze before it could escape. At that point, the aquifer would cap itself until the pressure built again. After the water escaped, the hydrostatic pressure needed to keep the pore space open in the aquifer would have dropped and the area above would collapse to produce the chaotic terrain.

Though catastrophic flooding is an attractive solution to how these channels could have formed under current conditions on Mars, it creates other mysteries. For example, considering that enormous amounts of water ran down these channels and the surface materials eroded to form the channels, there is little direct evidence of what happened to either the sediment or the water. It is clear that the enormous amount of water released during these floods and the sediment it carried would ultimately pool in the northern lowland plains to form huge lakes or small seas. Where are these bodies of water and do we see evidence of the sediments carried into them?

Fretted Channels
In contrast to the other channels and the valleys, there are wide, flat-floored, steep-walled sinuous troughs, called fretted channels, that have also developed on Mars. These channels exist only in association with the fretted terrain along the southern highlands boundary with the northern plains (see Figure 31). The slopes in this region appear to have been extensively eroded back into cliffs by mass wasting processes. Why such extensive erosion of this type should be isolated to this region is a mystery.

The fretted channels probably formed initially by either surface runoff or groundwater processes, as did the other valleys on Mars. Subsequent to their formation they were enlarged by mass wasting processes. Some of these valleys contain streamlined landforms, such as teardrop-shaped islands. Smaller features such as scours, cataracts, and jagged depressions are completely lacking. These smaller features may be buried by debris that eroded from the walls and now covers the floors of the valleys.

Grooves and striae have formed in the surface of the materials on the floors of these channels (Figure 72). Grooves and striae commonly parallel the walls in the middle of the valleys and curl downslope near the wall. It is almost as if the materials had flowed away from the walls and down the valleys. These patterns result from erosion of rock and soil from the valley walls probably assisted by ice in the debris, similar to terrestrial rock glaciers.

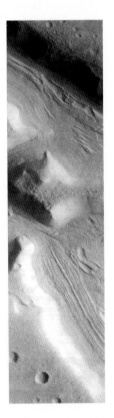

Figure 72. Broad, flat-floored, sinuous channels, called fretted channels, are part of a complex transition region, called fretted terrain. This terrain is found between the ancient heavily cratered highlands to the south and the sparsely cratered lowlands to the north. In this high-resolution picture, the grooved and striated surface of the floor materials found on the floor of one of these fretted valleys can be seen. These striations appear to represent material that has been shed from the smooth canyon walls and was subsequently modified by wind. (Courtesy NASA/Jet Propulsion Laboratory/Malin Space Science Systems, Incorporated) (PIA02075)

Other features on Mars also appear to have been cut by flowing water. Some of these features are enormous and have characteristics shown by terrestrial channels that formed in huge floods. The Martian features appear to have been cut by water fed by underground sources.

Lakes, Seas, and Oceans

With the abundant evidence that a substantial amount of water has flowed across the surface of Mars in the past, it is clear that there must be places where this water pooled before soaking into the ground, escaping into the atmosphere, freezing, or being buried by sedimentary or volcanic debris.

There are many places on Mars, particularly inside large craters and canyons, that contain layered deposits thought to be sediment carried by water that flowed through these features (Figure 73). Some of the craters and canyons also contain channels and valleys that either traverse or terminate in them, and they could be the location of paleolakes that formed while the channels and valleys were active and carrying surface water. For such lakes to develop these deposits, the lakes must have persisted as standing water for several thousand years. This has potential implications for climate history. As Victor Baker observed, "Standing water in lakes for such time-scales requires a major change in climate from present-day conditions, although ice-covered lakes could persist for considerable time in the current conditions or for those associated with the range of variation in Mars' orbital parameters."

On a much grander scale, most scientists expect that the largest outflow channels would have delivered their immense discharge to the northern plains. Their floodwater would have pooled in the Chryse Basin and in the lowest parts of the northern lowlands to form seas or small oceans. The sediments carried by the floodwaters from the huge outflow channels would have settled to the bottom of these bodies of water. For a time these bodies of water would have remained liquid, though it is expected that not long after the floods, the water would have frozen. This initially would have produced icy sediments, capped by clean ice. The ice would have sublimed quickly into the atmosphere and been redeposited at the poles, leaving behind ice-rich sediments capped by a layer of ice-poor sedimentary lag. This cap of dry debris would have effectively insulated the sediment and ice beneath and substantially slowed the rate of sublima-

tion of the ice. Consequently, a vast deposit of ice-rich sediment capped by ice-poor lag could have covered the northern lowlands.

Such short-lived ancient seas could explain the remarkable flatness and great variety of landforms found on the northern plains. Many of these features are suggestive of the action of a considerable amount of ground ice, and others are suggestive of shorelines. Tim Parker of Jet Propulsion Laboratory and Jim Head at Brown University and their colleagues searched for evidence of these ancient seas in the northern lowlands. Parker mapped ridges on the northern lowland plains and suggested that they may be strand lines or beaches of the shores of ancient bodies of water (Figure 74). Using the high-resolution topographic data from Mars Global Surveyor, Head analyzed the region mapped by Parker and suggested that there are "two distinctive basins in the northern lowlands, the circular Utopia basin of probable impact origin and the irregularly shaped North Polar basin." He calculated that it would take about 50,000 km^3 (10,700 cubic miles) of water to fill these basins to the strand line proposed by Parker (Figure 75). Considering the estimated water discharge rates from each channel-forming flood event, he estimated that "At least 40 events would be required to fill the northern lowlands to the level of contact 2 (i.e., the most probable strand line location)." But some of the water would probably disappear due to sublimation and soaking in between events and that to fill the basin "very high volumes, multiple discharge events through each channel, or a narrow range of emplacement times would be required."

However, other scientists argue that these ridges are not strand lines at all. They have found that they cross over hills and that vast regions in the northern lowlands appear to be covered by lava flows

---

Figure 73. Layered deposits are common in impact craters on Mars. Hundreds of layers of similar thickness are found exposed in the floor of this impact crater 64 km (40 miles) wide in western Arabia Terra. These layers provide a record of repeated episodes of sediment deposition and erosion. The layers toward the center lie nearly horizontal, but those near the edge of the crater are slightly tilted toward the crater center. These relationships suggest that the sediment that formed the layers was deposited from above, either settling out of the Martian atmosphere or out of water that partially filled this crater as a lake. (Courtesy NASA/ Jet Propulsion Laboratory/Malin Space Science Systems, Incorporated) (PIA02842)

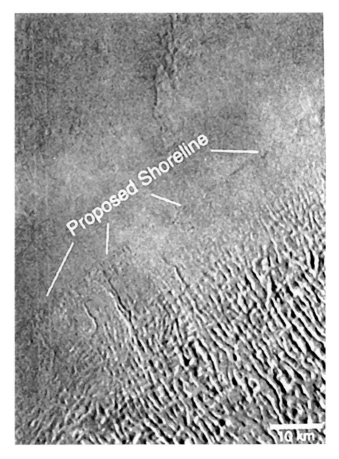

Figure 74. This Viking picture was one used to propose a shoreline around the North Pole Basin. High-resolution Mars Observer camera pictures generally do not show the finer scale details expected if these are shorelines, but because the scale of features observed in the high-resolution images is so small, these features may not be present now due to erosion or deposition even if the shoreline had been there in the ancient past. (Courtesy NASA, Viking) (129A32)

and not sedimentary plains. They suggest that sediments deposited in the northern plains by these great outflows of water have been buried by lava flows.

In spite of the current controversy over these features, the channels and valleys of Mars provide convincing evidence that water has played a substantial role in the development of the Martian surface. Where this water currently resides is a matter of great importance to

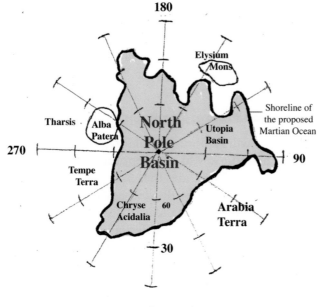

180

Elysium Mons

Tharsis

Alba Patera

North Pole Basin

Utopia Basin

Shoreline of the proposed Martian Ocean

270

Tempe Terra

90

Chryse Acidalia

60

Arabia Terra

30

0

Figure 75. Sketch map of the outline of the proposed shorelines of an ancient sea that may have occupied the northern lowlands of Mars. There may have been several episodes of flooding that formed seas in the northern lowlands, and this map shows the proposed shoreline from but one of those bodies of water. These seas may have formed as a result of the catastrophic release of subsurface water, mainly from the Valles Marineris region. The heavy black line traces the boundary and yellow shading indicates the location of a possible high stand of water. The elevation of the proposed shoreline varies little from the −4 km (−2.4 miles) elevation level, consistent with what would be expected for a water line. (Modified from T. S. Parker, R. S. Saunders, and D. M. Schneeberger, Transitional morphology in the West Deuteronilus Mensae Region of Mars: Implications for the modification of the lowland/upland boundary, *Icarus* 82:111–145, 1989; and J. W. Head III, H. Hiesinger, M. A. Ivanov, M. A. Kreslavsky, S. Pratt, and B. J. Thomson, Possible ancient oceans on Mars, *Science* 286:2134–2137, 1999)

the evolution of Mars as well as to its exploration and is the subject of vigorous debate. As we will see in the next chapter, there is evidence that some of this water resides in the regolith, frozen as ground ice, and some is in the polar caps or resides as a thin frost layer on the surface at the winter pole.

The Martian Deep Freeze: The Polar Regions

In this section, we will explore the polar regions of Mars. These fascinating places hold the key to an important part of the evolution of the Martian surface. They are the cold traps for an enormous amount of frozen water, carbon dioxide, and other volatiles. Volatiles have shaped the surface in the polar regions as well as periodically affecting the entire surface of Mars. Periodic changes in climate control the deposition, erosion, and stability of volatiles in this region. The details of these climate changes are discussed in chapter 6.

As on Earth, the poles of Mars receive less sunlight and as a result are colder than its equator. Mars is inclined at about 25° on its axis of rotation, similar to Earth's 23.5° tilt, and as a result has seasons. But Mars is in a highly elliptical orbit that changes its distance from the Sun considerably during the year. This makes the duration and temperatures of each season different from the others (e.g., winters at the South Pole are much colder than winters at the North Pole). Remarkably, the situation changes periodically. The gravitational pull from other solar system bodies tugs at the orbit and axis of Mars, continuously changing its characteristics. Changes in astronomical characteristics have dramatic effects on solar heating of the surface, especially in the polar regions. Heating controls the stability of frozen volatiles stored in the polar regions of Mars and also when the season of maximum dust storm activity occurs. Dust storms are particularly important because they provide airborne dust grains for nucleation sites for condensation of frost from the atmosphere. Thick dust clouds can shade the surface from solar heating.

There is considerable evidence that an enormous inventory of water and carbon dioxide is stored at the polar regions of Mars. The most obvious storage site is in its polar ice caps, though they only contain a small portion of the total inventory of Mars. Most evidence suggests that an enormous reservoir of ground ice is locked up in the regolith at high latitudes. Steve Clifford summarized the general view of most Mars scientists: "evidence suggests that Mars is water-rich and may store the equivalent of a global ocean of water of about 0.5–1 km [0.3–0.6 mile] deep as ground ice and ground water within its crust." On Mars, winter temperatures fall below the frost point of both water and carbon dioxide above about 80° latitude. The cold temperatures allow carbon dioxide and water vapor from the atmosphere to condense on the Martian surface. Above about 30° latitude, average surface temperatures are too warm for surface frosts, but are low enough for ground ice to be stable. A huge amount of volatile components must also be trapped on mineral grains. Under warm, wet conditions, some mineral grains readily absorb volatiles either on their surfaces or by locking into their structures. Releasing these volatiles back into the Martian atmosphere is difficult and requires unusual conditions.

In the lower latitudes, conditions are very different. Surface and subsurface temperatures are always too warm for water ice or carbon dioxide to be stable. Sublimation and diffusion processes quickly desiccate the upper part of the Martian regolith in this region, leaving it dry down to a depth of several hundred meters. Volatiles that evaporate from this region quickly migrate to the planet's cold traps, the polar regions. Although the loss rate of such volatiles from the regolith is easily calculated, recent observations of small, young channels suggest that Mars is a planet with an active hydrologic cycle. It is constantly being recharged, even under current conditions. Because the polar regions are the planet's cold traps, they must play an important role in this cycle.

Surface Ice: Ice Caps and Layered Deposits

Ice and frost are found on the surface at both poles as large seasonal caps that come and go annually and as small residual polar caps that generally remain all year round. The polar ice caps provide the only direct observational evidence of ice on Mars, though at high latitudes landforms that resemble terrestrial glacial and periglacial (those caused by ground ice) features indicate the presence of ground ice.

In an annual seasonal cycle beginning in fall, temperatures drop, the polar hood (a thick atmospheric haze) forms over the winter polar region, and water and carbon dioxide snows and frosts accumulate on the surface. By winter the polar hood dissipates, revealing the polar caps. As spring dawns and the polar region begins to warm, seasonal ice deposited during the winter systematically begins to retreat poleward (Figure 76). The retreat of this ice is quite rapid, suggesting that it is a thin deposit of mainly carbon dioxide frost. Its rapid retreat often leaves behind short-lived large patches of frost. These patches are generally on slopes that face away from the sun and remain for only a few days after the disappearance of the surrounding ice. Some of these patches, such as the Mountains of Mitchell (which are not mountains at all) in the southern polar region, are large enough to have been mapped by early astronomers.

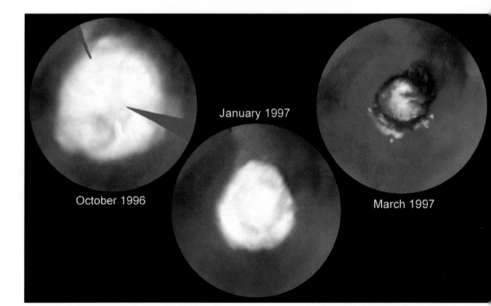

Figure 76. Polar projection view of the north polar ice cap of Mars during three different seasons. Differences in seasonal extent of the polar cap are clearly shown. Each view is a mosaic containing three images, collected by Hubble Space Telescope at a different time of the day, assembled to show the entire pole in daylight. The image on the left shows the North Pole region in late winter/early spring when the seasonal cap extends south to about 60°N latitude. The center image shows that by midspring the cap has retreated to 70°N latitude. The image on the right was taken during early Martian summer and shows only the north residual ice cap. Chasma Borealis, the huge canyon that cut into the polar terrain, is clearly visible in this summertime image. (Courtesy NASA/Space Telescope Science Institute) (PIA01247)

As the ice continues to recede, the sharp outlines of the small permanent caps are eventually discernable.

The two permanent caps appear to have some important differences from one another. The permanent cap in the north (Figure 77) is much larger than the South Pole cap (Figure 78), about 1,000 km (600 miles) across compared with about 350 km (210 miles) across, respectively. The northern cap is thought to be composed of water ice, and the southern cap of a mixture of water ice and carbon dioxide frost. The summer temperatures push the surface to about −68°C (−90°F), well above the frost point of carbon dioxide, but below the frost point of water ice. As a result high summer temperatures would drive off the carbon dioxide and leave the water ice. In contrast, temperatures during southern summer at the South Pole fail to exceed −125°C (−193°F), below the frost point of both water and carbon dioxide ices. Because of its lower temperature, the southern cap should act as a cold trap for both carbon dioxide and water ice year-round. As a result, most scientists think that the

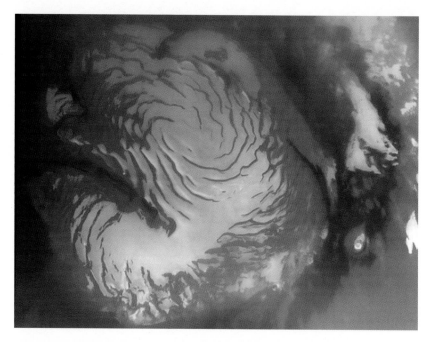

Figure 77. The north permanent polar ice cap. The swirling form is probably caused by preferential removal of frost on south-facing slopes. (Courtesy U.S. Geological Survey)

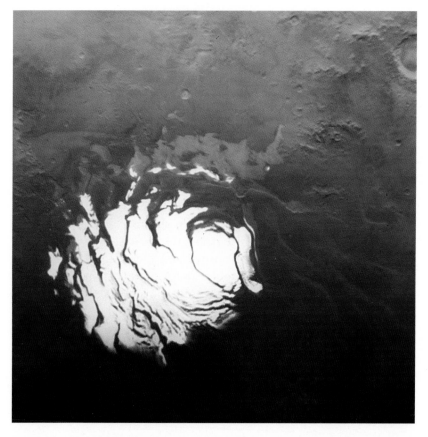

Figure 78. View of the south permanent polar ice cap. Layered terrain, with smooth, lobate, outward-facing escarpments, covers most of the area around the cap. (Courtesy U.S. Geological Survey)

South Pole cap is mainly carbon dioxide ice with minor amounts of water ice.

The temperatures at the poles are influenced by global dust storm activity. Currently, global storm activity occurs mainly during southern summer when Mars is close to the Sun in its orbit. These intense storms produce a dust-laden atmosphere that shades the southern cap from the peak period of solar heating during southern summer, preventing its temperature from climbing. At the same time in the north, where it is winter, these storms fill the atmosphere with dust that acts as nucleation sites for condensation of large amounts of carbon dioxide and water. These ice-coated dust grains settle to the surface in a deposit of dusty ice. The high dust content of the deposit decreases its reflectivity and as a result it readily absorbs sunlight. This increases its temperature, driving off the carbon dioxide in the spring and summer while leaving much of the water frost. During southern winter, when dust storm activity is at a minimum, frost deposits are cleaner and as a result cooler because they reflect more sunlight than their northern counterpart. These effects result in a maximum temperature difference between the northern cap and southern cap of about 45°C (81°F). Changes in obliquity reset the timing of dust storms so that over time they occur throughout the year.

Layered Deposits

Both permanent ice caps sit on much larger finely layered deposits of dust and frost (Figure 79). These layered deposits contain only a few impact craters and appear to be young, at most only a few tens to a few hundreds of millions of years old. As with the polar ice caps, these deposits are formed by deposition of frost and dust from the atmosphere. Individual layers are thin, some less than the limit of resolution of orbiter pictures (Figure 80). Some can be traced for several hundred kilometers, others for only several tens of kilometers. In a few locations, layers terminate against other layers. On Earth this type of pattern is caused by multiple episodes of deposition and erosion in a region.

The layered appearance is probably due to differences in the relative proportion of dust and ice in each layer. Such layering suggests cyclic changes in the conditions that formed them. Most likely, the layers record surface conditions that oscillate between deposition of dusty ice and relatively clean ice. The truncation of layers against other layers is another matter. This probably indicates that the periods of deposition have been broken by periods of erosion.

A series of arcuate-shaped valleys and low escarpments is incised into these deposits at about 50-km (30-mile) intervals. These valleys curve outward from the poles and are outlined by the bright ice of the polar caps, giving them a spiral appearance. In the north polar region the valleys curve counterclockwise (eastward) and in the south they curve clockwise (westward). These valleys are thought to be the result of the preferential removal of frost from the warmer westward-facing slopes caused by solar radiation striking them more directly. Because the western slopes get more direct sunlight they

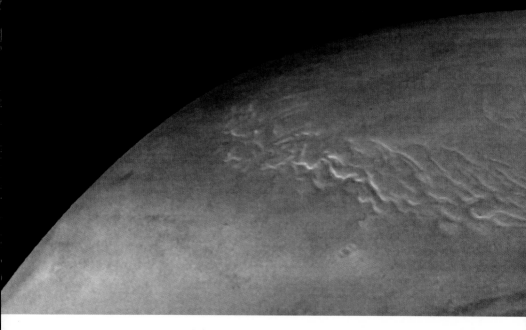

Figure 79. Layered deposits dominate this view of the north polar region of Mars. This terrain is believed to be composed of ice and dust deposited over millions of years. The swirled pattern is caused by channels eroded into this deposit. (Courtesy NASA/Jet Propulsion Laboratory/Malin Space Science Systems, Incorporated) (PIA01471)

are warmer and as a result the frost sublimes more readily from them. The ices that sublime are quickly redeposited on nearby flat surfaces that receive less solar radiation and are cooler.

At a regional scale, the surface of the layered deposits is generally smooth and gently rolling, with fine linear fluting sculpted by the wind. At a finer scale the surface is highly textured, with a host of mounds, cracks, and ridges. Most of these are several meters to several tens of meters across and probably are a result of repeated sublimation and deposition on these surfaces.

What is the relationship between the polar ice caps and these ice-rich layered deposits? Most likely, the ice caps are simply the bright surficial ice units of the layered deposits. Clifford has observed that "The two units (permanent ice caps and layered deposits) are distinguished by their temperatures and albedos, but the residual caps may simply be the most recent layer in the process of forming. In as much as the layers almost certainly represent geographic and temporal variations in the mass balance of ice and dust, the present-day residual caps are most likely comparable to only specific parts of the layered terrain."

The relative sizes of the ice caps have an inverse relationship to the size of the layered deposits. Though the southern ice cap is

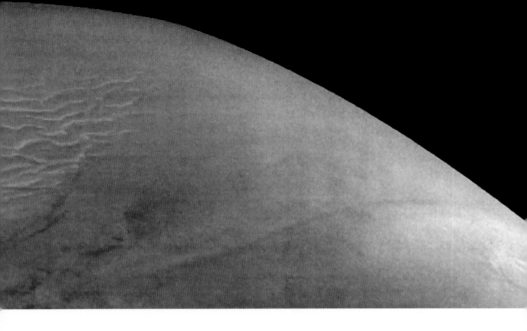

smaller than the northern ice cap, the southern layered deposits are much larger than the northern layered deposits. These deposits are comparable in relief, about 3 km (1.8 miles), and, considering their areas and the flexure of the basal surface due to their weight, are about a factor of two different in volume from one another. The southern deposits contain about 2–3 million km^3 (422,000–657,000 cubic miles) of materials, and the northern deposits about 1.5 million km^3 (328,500 cubic miles) of ice and dust. This difference in size may be the result of climate changes caused by quasiperiodic changes in the motion of Mars.

The young age of the layered deposits and the evidence of periodic swings in climate suggest to scientists that these deposits should have older counterparts, laid down earlier in Martian history when conditions were right for deposition and later eroded when conditions were right for removal. If this is true, do remnants of these old layered deposits still exist on Mars? The answer may be yes. They are found in the south polar region as etched and pitted deposits. These deposits are found around the edge of the layered deposits and also are thought to be composed of ice and dust. They were probably laid down earlier in the history of Mars when the climate favored deposition of large amounts of dust and ice. In the cold regions on Earth,

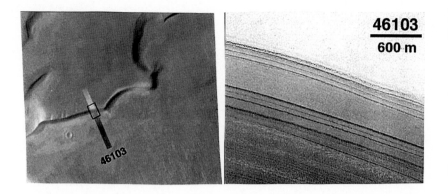

Figure 80. The Martian polar regions are covered by large areas of layered deposits. The high-resolution view on the right, of the slopes along the edge of the layered terrain surrounding the North Pole, shows that the layers have different thickness and different physical expression. Some of the layers form steeper slopes than others, suggesting that they are more resistant to erosion. All layers have rough texture, indicative of erosion. On the left is a regional-scale view containing the area in the high-resolution picture (in the box). (Courtesy NASA/Jet Propulsion Laboratory/Malin Space Science Systems, Incorporated) (PIA01479)

etching and pitting of the surface is common, caused by the sublimation of ground ice. The Martian etched and pitted deposits appear to have undergone several episodes of erosion and deposition. These episodes probably reflect periodic changes in climate. Erosion during these episodes did not affect the underlying ancient cratered terrain. This indicates that the etched and pitted deposits are much less resistant to erosion than the underlying terrain, as would be expected if the deposits are composed of dust cemented by ice and the ancient terrain is composed of igneous rock.

Periglacial and Glacial Landforms

On Earth unique landforms produced by ice are found in its coldest regions (in its polar regions and at high elevations). Some landforms are glacial, produced by massive ice sheets that scour the surface and build deposits from the debris that they erode. Other landforms are periglacial and are produced by freezing of the ground and the water it contains. The development of these landforms generally involves ablation, ice-assisted mass wasting, and slow creep (solid-state flow of icy surface materials). Presumably, the Martian landforms are produced in the same way as the terrestrial landforms. Although interpretation of the mechanism of formation of specific features is sometimes debated, scientists generally agree that periglacial processes (involving fracturing by freezing and thawing of water and related processes) have been important to the development of the surface of Mars.

In the southern highlands, particularly in the mountainous regions around Hellas and Argyre impact basins, the landscapes have been proposed to be an intricate assemblage of glacial landforms. Jeff Kargel of the U.S. Geological Survey and Robert Strom of the University of Arizona have championed this view and feel that they have identified such glacial features as horns, cirques, aretes, grooves, V-shaped valleys, rock glaciers, or debris-covered glaciers in the Charitum Montes region south of Argyre Basin. The uplands near Hellas show features interpreted to represent very recent glacial flow: lobate debris aprons, crevasselike fracture concentrations, and medial moraines having glacierlike patterns of tributary convergence and downvalley flow. If these landforms are truly formed by glaciers, then they need a large snow accumulation area to continue to be fed ice. That requires atmospheric transport of water to the site and a net surplus of input from snow to offset losses by sublimation and meltwater runoff. However, the large accumulation area might not be required if the glacier is covered with debris, reducing the ablation rate of its ice.

In addition, Kargel and Strom have also suggested that the floor of Argyre contains numerous examples of other types of glacial landforms such as eskers, kettles, drumlins, moraines, and outwash plains (Figure 81). Farther south, eskerlike ridges are also found in the Dorsa Agrentea region (75–80°S). These are thought by some scientists to have been produced by the melt-back of ice deposits derived from an extensive south polar ice cap in the middle of the planet's history. It has been suggested that drainage from this ice cap flowed through the Argyre Basin, carving prominent channels and creating a temporary lake (and perhaps a glacier).

In the north, the polar layered deposits sit on sparsely cratered lowland plains that contain a great variety of peculiar landforms. Many of these landforms look like extraterrestrial counterparts of landforms that have developed on Earth through the action of ground

Figure 81. These eskerlike features found in the Argyre Basin are sharp-crested ridges as much as 200 km (120 miles) long and 2 km (1.2 miles) wide. These features have also been interpreted to be other types of landforms, but together with the associated glacial features suggest that Argyre might have been the location of considerable glaciation. (Courtesy NASA, Viking) (567A33)

ice and permafrost (Figure 82). There are arrays of closed depressions, polygonal fractures, mottling or patterns of stipples, moats around low hills, and pedestal-shaped impact crater ejecta, most of which are probably periglacial features produced by freezing of water-soaked ground. The water is thought to have come from the repeated flooding of the plains from the outflow channels. These features often cluster into provinces. This probably reflects regional conditions at the time of formation of these landforms.

Two types of polygonal-patterned ground are found on the lowland plains that are imprinted by shallow fractures. There are fields

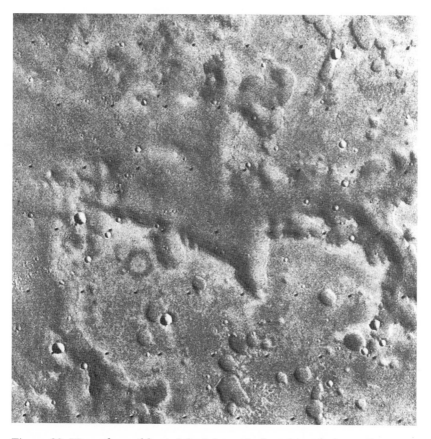

Figure 82. View of possible periglacial terrain found in southern Chryse Planitia. In this picture, small, irregular depressions have formed in the materials that make up the plateau. These small depressions appear to have enlarged and merged to form valleys as the plateau was progressively destroyed. The depressions are similar in size and form to alases, which develop on Earth by collapse of ground containing ice and water. The scalloped edge of the plateau may indicate cliff retreat enhanced by coalescing depressions. The image is about 40 km (24 miles) across. (Courtesy NASA, Viking) (8A74)

of small polygons in the polar regions of Mars (Figure 83) that are about the same size and shape as polygonal ground found in the polar regions of Earth. The small terrestrial polygons are about 5–10 m (16–32 feet) across and are produced by stresses from the repeated freezing and thawing of water in frozen soils. In addition to the systems of small polygons, giant polygonal fractures are found on Mars. These features, concentrated in the Acidalia Planitia and Utopia

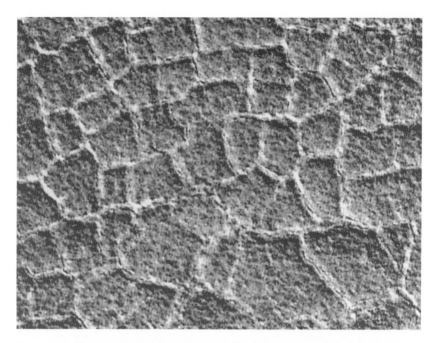

Figure 83. These polygonal patterns, each a kilometer or less across, are found in the floor materials of some impact craters on the northern plains. These features are very similar in size and shape to features common in the polar regions on Earth and usually indicate the effects of freeze-thaw cycles of subsurface ice. (Courtesy NASA/Jet Propulsion Laboratory/ Malin Space Science Systems, Incorporated) (PIA02072)

Planitia regions, superficially resemble patterned ground in polar regions of Earth (Figure 84). However, these Martian polygons are about two orders of magnitude larger than the largest patterned ground on Earth. Stress systems produced by this mechanism are incapable of producing polygons on the scale of the Martian features. Because of this, scientists think that it is likely that the large Martian polygons are the work of a regional-scale geologic process, such as regional doming.

Extensive regions of "patterned plains" are found on the northern lowlands (Figure 85). These plains are composed of two major rock layers, a layer that forms smooth plains that overlay an erosion-resistant layer beneath. The smooth plains appear to be partly stripped away and could be composed of such easily eroded materials as windblown or flood sediments.

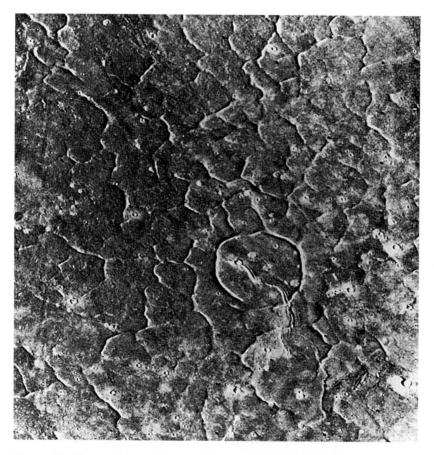

Figure 84. Giant polygonal fractures are common in the northern plains in the Acidalia Planitia region. These cracks are thought to have been caused by regional stresses. The scene is about 53 km (32 miles) across. (Courtesy NASA, Viking) (32A18)

North of 50°N, there are extensive regions of dark-mottled plains (Figure 86). The mottled appearance of these plains is the result of the marked contrast between their dark surface and bright material of superimposed impact craters. Why the plains and craters are composed of materials of such different brightness is unknown. Some researchers have suggested that the high reflectivity of the crater materials may be due to quarrying of bright material from the subsurface; others contend that it is the result of bright dust trapped by the rough-textured blanket of debris ejected from the craters.

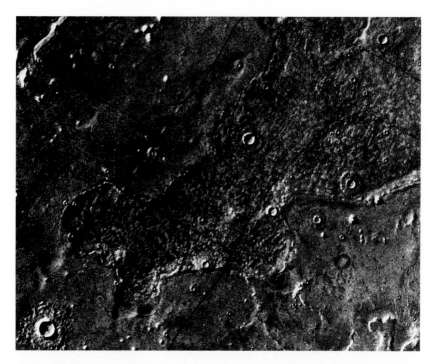

Figure 85. Patterned plains in northeast Acidalia Planitia. The origin of
this terrain is not known, but it may have been formed by stripping of
sedimentary materials that were deposited in the northern lowlands by
huge floods. The scene is 130 km (78 miles) across. The straight dark lines
are boundaries between individual pictures in the photomosaic. (Courtesy
NASA, Viking) (211-5066)

Extensive plains found northwest of Elysium contain irregular arrays of closely spaced hills and hollows that resemble terrain found
in cold, volatile-rich regions on Earth (Figure 87). In the Elysium region, large, confined outflow channels connect with narrow, meandering troughs that cross these disordered plains, as they are called.
These troughs terminate in close-spaced ridges having a pattern
reminiscent of terrestrial distributary channels. From the degree of
their complexity, it is clear that these plains have had a complicated
erosion and deposition history. Most scientists agree that the disordered plains were originally formed by ice or running water fed
from the Elysium area, indicating that there was and may still be an
abundance of subsurface water and ice in this region of Mars.

It is clear from landforms found in the high-latitude regions of
Mars that there are (or were) huge amounts of volatile materials on

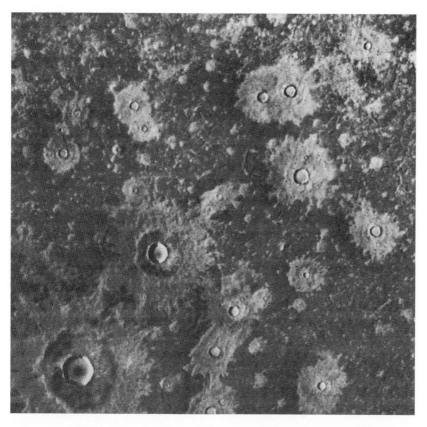

Figure 86. Mottled plains in Acidalia Planitia. The dark plains contrast sharply with the light-colored impact crater ejecta. Most of the craters have morphology unique to northern regions suggestive of subsurface ice. The scene is 245 km (205 miles) across. (Courtesy NASA, Viking) (673B30)

the surface and stored in the subsurface. These volatiles are critical to the evolution of Mars and its climate. In particular, if the volatiles in the subsurface are still there, what does it take to release them and what would the effects be? Would their release produce a milder climate? How long would it last? If the volatiles are no longer trapped in the subsurface, then where did they go and why? Some of these questions are addressed in chapter 6, but most are still unanswered.

Interpreting the Cratering Record

Impact craters are found on nearly every solid body in our solar system. Through the study of these craters, we have found that their

Figure 87. Disordered plains several hundred kilometers west of Elysium containing many closely spaced hills and hollows like those found in permafrost terrain on Earth. The channel at the bottom of the picture originated near Elysium. The picture is about 120 km (72 miles) across. (Courtesy NASA, Viking) (9B50)

formation has been an important factor in the surface evolution of all planets. We have also found that they provide a valuable tool for estimating surface age, discerning physical and chemical properties of crustal materials, and detecting processes that have affected the surface.

Impact craters are found on every terrain on Mars, except the polar ice caps. In marked contrast, impact craters are rare on Earth, although, like the other planets, Earth surely must have been bombarded throughout its history. We know that craters larger than about 1 km (0.6 mile) in diameter have been formed every few hundred thousand years on Earth. Those larger than 10 km (6 miles) have been formed every several tens of millions of years, and those greater than 100 km (60 miles) about every billion years. Why have only a small number of these craters been found? Earth's dynamic surface and tectonic environment quickly erase all traces of most of them. The same holds true for all other landforms on Earth. Because of its dynamic surface environment, landforms on Earth last no longer than a few hundred thousand years. In contrast, erosion and resurfacing rates on most other planets are much lower than on Earth, allowing impact craters and other surface features to survive much longer.

Before close-up pictures were taken of Mars, scientists could not agree about whether impact craters would be found on its surface. Some scientists felt that, as on Earth, active surface processes would erase craters soon after they formed. Others assumed that, as on the Moon, the erosion rate would be low and impact craters would be common. A single strip of pictures taken by Mariner 4 settled this argument once and for all. The pictures showed heavily cratered terrain like that on the Moon, but more subdued. The size frequency (the number of craters of particular diameter range) was also similar, though not identical with that of craters on the Moon. For crater experts, this was good news, but for those hoping for an Earth-like Mars, it was disappointing.

As exploration continued and high-resolution pictures were sent back of all of Mars, some surprising and important differences were found between Martian impact craters and craters on other solar system bodies. In the next section, we will explore the cratering record on Mars and what make it so distinct from that of other planets. As a prelude to this discussion, we will start with how craters are formed and how they are used to provide information about the surface they penetrate.

How to Make an Impact Crater

Natural impact craters are created by a sudden burst of kinetic energy released when space debris, such as an asteroid or comet, plunges into a planetary surface. Typically these bodies strike the surface at a speed of a few kilometers per second, but may make this plunge at up to 70 km (42 miles)/sec. This pulse of energy radiates outward through the target rock as a strong shock wave at velocities ranging from about 6 to 20 km (3.6 to 12 miles)/sec, depending on the strength of the shock wave. Shock pressures are so high close to the point of impact that they vaporize or melt both nearby target rock as well as the asteroid or comet. Farther away, the target rock remains solid though it may be crushed, deformed, and ejected. As the shock wave expands outward, target materials are instantaneously compressed and accelerated away from the point of impact. Immediately behind the shock wave pressures drop to near zero, and the outward velocity of the target material drops to about one-fifth of its peak speed. As a result of this outward motion, the target rock immediately behind the shock wave is deflected laterally and upward. If these materials are moving fast enough, they will be ripped from their places and ejected from the crater. Crater excavation continues until this movement can no longer overcome the strength of the rock and force of gravity. The shock wave continues outward, rapidly weakening and slowing as it expands, because of both spreading and loss of energy used as it crushes and heats the target rock.

Though this general scheme is the same for all craters, their final shape is dependent on their size. Small craters, generally less than a few kilometers in diameter, are bowl-shaped pits with raised rims. This shape reflects the hemispherical shape of the shock wave frozen in the rock. Their depth is typically about one-fifth of their diameter. Craters larger than several kilometers in diameter typically form with flat floors and central mountain peaks. The occurrence and number of peak rings is related to the size of the crater and thickness of the planet's lithosphere at the time of impact and is controlled by the strength of the gravity field of the planet. These interior features are probably the result of elastic rebound of the shock-compressed target material beneath the craters. This rebound motion moves the material beneath the crater inward and upward, pushing up central peaks and peak rings. These rebound forces also occur in small craters but are usually too small to overcome the strength of the target rock. Though most of the rebound occurs during cratering, stress can be stored in the compressed target materials and rebound slowly over an extended period. The transition diameter from small, simple, bowl-shaped craters to complex flat-floored craters is mainly controlled by the relative strength of the gravity field. Consequently, complex features are initiated at smaller sizes on more massive planets.

For most impact craters found in our solar system, the rock thrown out of the crater, called ejecta, travels in ballistic trajectories, much like the path of artillery shells. This material falls back to the surface and is deposited around the crater. Closest to the crater, the ejecta forms a relatively thick, continuous, hummocky deposit that rapidly thins outward into a discontinuous, thin layer of material. Farther out, the discontinuous ejecta breaks up into rays, clusters, and chains of "secondary" craters produced by far-flung debris.

The ejecta deposits of many Martian craters are remarkably different from those found around craters on other planets. Many Martian crater ejecta deposits show evidence of flow across the surface during their emplacement. Why and how this flow occurred is still a matter of intense controversy. Most scientists think that the flow resulted from the fluidizing effects of abundant volatiles in the ejecta. Some of these scientists liken the flow to gas-driven ash flows from volcanoes or massive landslides. Others suggest that this type of ejecta flowed across the surface more like a slurry of wet debris during its emplacement. But other scientists, notably Pete Schultz at Brown University and his colleague Oliver Barnouin-Jha at the Applied Physics Laboratory, have suggested that this type of ejecta is dominantly the result of the effects of turbulence in the thin Martian atmosphere.

The material ejected during crater excavation is dominantly composed of target rock, though small amounts of the impacting body are also commonly included in these materials. The ejection of these materials systematically follows the shock wave as it radiates outward. For ballistic-type craters, deposition of these materials on the surface faithfully follows the order of their ejection and results in ejecta deposits that are an overturned version of the sequence of rock layers in the target. This phenomenon provides a simple, free (unlike core drilling), and predictable means of sampling subsurface strata. Apollo astronauts used the phenomenon to collect subsurface samples from the Moon. Providing care is taken in sample selection, the same technique could prove valuable for future sample-collection missions to Mars. Or will it?

Fluidization may complicate such systematic emplacement of ejecta. Turbulence within the flowing ejecta may cause considerable mixing of its components. Depending on the degree of mixing, the orderly inversion of the stratigraphic sequence of the target materials usually seen in ballistic ejecta deposits could be substantially disrupted. This disruption could seriously complicate sampling strategies for any Martian sample-collection missions.

What Impact Crater Populations Tell Us

Nearly every solid planetary surface in our solar system has been peppered with impact craters. These craters are created by the collision of asteroidal and cometary debris with the planets. With time, these craters accumulate to form populations, all with similar characteristics initially. Changes in the characteristics of these populations can used to detect geologic processes that shaped the surface, and the relative crater density is frequently used to estimate surface age. What makes this possible?

The relative age of a surface can be estimated by using crater density because the rain of projectiles from space has been continuous and uniform on each planet. Impact craters are mainly produced on a planet by the continuous sweeping up of debris left over from solar system formation. Debris generally travels at much greater velocities than the orbital motion of the planets, causing it to strike the planet from all directions. This means that if the surface of an entire planet has uniform age, it would also have uniform crater density.

As a result, in the simplest case, crater density is a function of relative age. From our studies of the solar system, we know that the frequency of collisions between debris and the planets has declined exponentially as this debris has been swept up. As a result, the density of craters on a surface only provides a relative age. The conversion of crater density ages from a relative scale to absolute values is difficult and tricky. For the Moon, this task was straightforward and required only the comparison of radiometric ages of samples collected by Apollo and Luna missions with crater densities measured for each area around the respective landing sites. This technique worked well enough that Laurence Soderblom, the author, and our colleagues at the U.S. Geological Survey in Flagstaff, Arizona, accurately predicted the age of the last several lunar landing sites before samples were collected. Age calibration studies also confirmed the suspicions of many lunar scientists that the impact rate on the Moon (and presumably for the rest of the inner solar system) was initially extremely high but plummeted about 3.9 billion years ago and has remained low and nearly constant since.

Unfortunately this technique of relative age calibration cannot yet work on Mars. Unlike the Moon, from which documented rock samples have been collected, no such samples have been collected from Mars. Consequently, on Mars only relative ages can be estimated reliably using crater data. In spite of this drawback, there are other ways to calibrate relative ages derived from crater density that may provide reasonably accurate absolute age estimates. Notably, Gerhard Neukum of the Institut für Weltraumensorik und Planetener-Kundung in Germany and his colleague William Hartmann of the Planetary Science Institute have extrapolated the lunar impact flux to Mars. They have assumed that the characteristics of impact flux are the same at both bodies but with different magnitudes. Underpinned by studies of the populations of asteroids and comets that could potentially hit Mars and the effects of the unique Martian environment, they have developed estimates of the differences in magnitude and have calculated absolute age calibration factors for crater densities on Mars. Although most scientists think that their estimates of the impact flux on Mars are reasonably accurate, the estimates remain untested until samples are collected.

Density is not the only characteristic of crater populations that provides information about the evolution of planetary surfaces. Deviation in the size frequency distribution of craters also provides such information. Many scientists believe that the size frequency distribution of craters has been nearly uniform with time, reflecting the results of collisional processes between asteroids, comets, and other debris during solar system evolution. Deviations from this crater size frequency distribution are generally the result of surface processes that have either selectively erased part of the crater population or added to it. For example, secondary impact craters (by debris ejected from primary craters) or volcanic craters can add considerably to the density of a crater population in an area, and erosion processes or flooding by lava flows can selectively erase small craters. These processes are identified by their effect on the size frequency distribution of the crater population.

Morphology of Martian Impact Craters

On all planets, crater shape changes in a systematic way with increasing size. Most morphologic changes with size are controlled by the strength of the planet's gravity field and target material properties, although on Mars some of these changes appear to be strongly influenced by the abundant volatiles in its crust.

As on other planets, craters on Mars smaller than 5 km (3.6 miles) in diameter are generally simple, bowl-shaped pits with raised rims. Craters larger than 8 km (4.8 miles) have complex interiors with flat floors, terraced rims, and central mountain peaks or peak rings. On Mars central peaks occur over a much greater size range of craters (1.5 to 250 km [0.9 to 150 miles]) and the peaks are comparatively larger across than on other planets. Unlike on any other planet, on Mars some craters have central peaks topped with a pit and some flat-floored craters contain a central pit instead of a central peak. Pits are thought to form by an explosive release of water trapped in the subsurface during crater formation.

At larger crater diameters, central peak formation gives way to peak ring formation. The smallest peak ring crater on the Moon is nearly 200 km (120 miles) in diameter, but the smallest peak ring crater on Mars is only about 45 km (27 miles) in diameter. As with other unique features on Mars, this may be due to the presence of abundant volatiles in the subsurface.

In small, bowl-shaped craters on Mars, slumps and slides are common on the interior wall. Most craters larger than about 5–10 km (3–6 miles) have developed terraces on their interior rim. These structures are similar to those in craters on other planets and are thought to be due to failure of interior slopes under their own weight. The onset of terracing occurs with slightly different crater sizes in different terrains on Mars, probably reflecting differences in strength of the rocks that form these terrains.

By far the most unique feature of Martian impact craters is their ejecta blankets. Most craters on Mars that are smaller than about a few kilometers and larger than 60 km (36 miles) in diameter are surrounded by radial ejecta morphology caused by ballistic transport. In contrast, ejecta surrounding craters between about 3–5 km (1.8–3 miles) and about 60 km (36 miles) in diameter typically shows evidence of flow during emplacement. A few craters show a combination of both fluidized and ballistic patterns.

In places, evidence of flow of the ejecta is clear. Where preexisting topography occurs, the ejecta appears to have flowed across the surface during its emplacement and was deflected around these obstacles. The surfaces of most of the deposits from fluidized ejecta craters are ridged and grooved radial to the parent crater, suggesting outward flow of the ejecta. In addition, the outer edges of these ejecta deposits terminate abruptly in a cliff. Often in map view this outer edge has an irregular, lobate flower shape. This pattern is difficult to explain by any other mode of transport.

Typically, the ejecta of these fluidized ejecta craters is composed of one or more complete or partial sheets of materials surrounding the crater. The number of layers these craters may have depends on the size and location of the crater. For example, large craters tend to have more layers than small craters, and double-layered ejecta craters are common on the northern lowland plains above 30–40°N but are rarely found elsewhere on Mars. Why these craters should have layered ejecta is unknown. Most scientists think that the individual layers mark changes in the crustal material properties with depth. Many scientists think that changes in properties could be a result of changes in volatile content with depth.

To help make sense of the many different variations of ejecta on Mars, Nadine Barlow at the University of Central Florida and her colleagues have laid out a nongeneric nomenclature that describes the craters of Mars, based on the morphology of their ejecta blankets (Figure 88). They call the craters surrounded by ejecta (which looks similar to the ejecta around craters on most other planets) single-layer (or single-lobe) radial ejecta craters. The ejecta of these craters probably was transported along ballistic trajectories. They note that the greatest diversity of ejecta types show layered patterns. These patterns are divided by how many layers of ejecta are present. Barlow indicated that "The single-layer, double-layer, and multiple-layer categories are modified by terms that describe the shape of the ejecta terminus. The layered ejecta patterns terminated by a distal ridge or rampart shall be modified by the term 'rampart.' Hence, single-layered ejecta patterns terminated in a distal ridge would be called a 'single-layer ejecta rampart.' Layered ejecta patterns that terminate in a concave slope will be modified by the term 'pancake.'" They modified the rampart and pancake terms further "by the adjectives 'sinuous' and 'circular,' describing the general sinuosity of the ejecta blanket."

There are other craters with layered ejecta patterns. Barlow and her colleagues call them pedestal craters because they sit perched above the surrounding terrain. These craters are found mainly in the polar regions. The ejecta deposit may extend out several crater

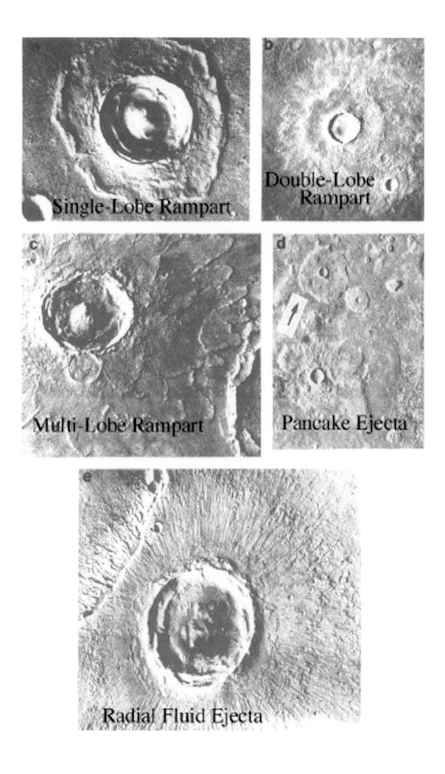

Single-Lobe Rampart

Double-Lobe Rampart

Multi-Lobe Rampart

Pancake Ejecta

Radial Fluid Ejecta

radii or terminate in less than one. This relationship has caused scientists to conclude that the shape of the ejecta blanket may, in part, be caused by erosion. Some scientists have suggested that during periods of erosion, blocky ejecta from these craters protects or "armors" the surrounding surface. As the surrounding region erodes away, the more-resistant ejecta deposit remains as a broad pancake-shaped pedestal. Erosion may continue to wear the feature away until only a mound with a summit pit remains, closely resembling a volcanic dome.

As with those found on other planets, impact craters are important landforms on Mars. The broad range in density and unique characteristics of the shapes of these craters provide information on age and crustal properties.

Figure 88. Examples of different types of Martian impact craters showing fluidized ejecta morphology: *a,* a crater 7 km (4.2 miles) in diameter with single-lobe (single-layer) ejecta with an outer rampart; *b,* a crater 16 km (9.6 miles) in diameter with double-lobe (double-layer) ejecta with outer ramparts; *c,* a crater 18 km (11 miles) in diameter with multiple-lobe (multiple-layer) ejecta with outer ramparts; *d,* a pancake ejecta crater 3 km (1.8 miles) in diameter; *e,* a crater 20 km (12 miles) in diameter with radial fluid ejecta. (Modified from R. G. Strom, S. K. Croft, and N. G. Barlow, The Martian impact cratering record, 383–423 *in* H. H. Kieffer, B. M. Jakosky, C. W. Snyder, and M. S. Matthews, eds., *Mars,* University of Arizona Press, 1992)

Chapter 6

The Atmosphere and Climate

When early astronomers first turned their telescopes toward Mars, they thought that they were looking at an Earth-like planet with an Earth-like atmosphere. They described whitish clouds floating across the planet's surface, seasonal hazes covering its polar regions, and small dust storms frequently erupting, a few of which grew to engulf the entire planet.

It was not until the beginning of the twentieth century that their impression of an Earth-like Mars began to change, brought about by a revolution in the development of modern scientific instruments. In the 1920s a series of attempts were made to measure the pressure of the Martian atmosphere using new instruments. However, attempts proved to be inaccurate because to work they required knowing the composition of the atmosphere. In the absence of compositional data, most scientists incorrectly assumed that the Martian atmosphere was Earth-like, composed mainly of nitrogen. A major breakthrough occurred in 1947 when Gerard Kuiper, using a newly developed infrared detector, discovered that carbon dioxide was a major component of the atmosphere. Soon other scientists began to discover other chemical components in the Martian atmosphere. The next major breakthrough occurred in the 1960s when scientists began sending scientific instruments to Mars aboard spacecraft.

Though the Viking landers were the first to make direct compositional and meteorological measurements, it was Mariner 9 that gave us the first long-duration weather satellite-like view of the Martian atmosphere. The Viking orbiters, Hubble Space Telescope, and Mars Global Surveyor extended these observations. Twenty-one years after

Viking, Mars Pathfinder significantly added to weather and meteorology data of Mars. The landers have been particularly important in collecting data about the atmosphere, not only carrying weather stations and composition-measuring devices to the surface, but also measuring the density, composition, and temperature in the atmosphere on their plunge to the surface.

Measurements from spacecraft have shown that the Martian atmosphere is indeed mainly composed of carbon dioxide and has only about one-hundredth the pressure of Earth's atmosphere. Near the surface, it ranges from a balmy 25°C (77°F) in the equatorial regions during midsummer to a frigid −125°C (−193°F) at the winter pole and by Earth's standards is dry. There is also mounting evidence that the atmosphere may not have always been like this.

In this chapter, we will explore what is known about the thin layer of gas surrounding Mars, its atmosphere. We will examine its current physical (weather and meteorology) and chemical state, its origin, what it may have been in the past, and how it has changed and why. In particular, we will explore one of the most fascinating aspects of the Martian atmosphere, its long-term behavior or climate history.

Martian Weather and Meteorology

Aside from the very limited atmospheric water and its effects on the heat balance, the main difference between the atmosphere of Mars and that of Earth is its very low surface pressure, which varies as much as 20 percent annually as a result of seasonal carbon dioxide condensation at the poles. Because of this, compared with other planets, the weather of Mars is remarkably similar to Earth's. Conway Leovy of the University of Washington, a Mars atmosphere expert and Viking Orbiter Imaging Team member, invites us to "Imagine a planet very much like the Earth, with similar size, rotation rate and inclination of rotation axis. Possessing an atmosphere and solid surface, but lacking oceans and dense clouds of water vapor. We might expect such a desert planet to be dominated by large variations in day-night and winter-summer weather. Dust storms would be common." He, of course, was talking about Mars. In the next two sections, we will discuss the Martian weather: the winds, clouds, and storms that are common in the Martian atmosphere.

Winds and Circulation Patterns

Remarkably, during each Martian year nearly 20 percent of the mass of the atmosphere migrates from pole to pole. This seasonal exchange is unique to Mars and is called condensation flow (see Figure 89). Condensation flow occurs because temperatures in the Martian polar regions typically hover near the frost point of carbon dioxide and water. As a result, small changes in temperature at the poles, such as occurs during the changing seasons, can cause huge amounts of volatiles to either freeze out of the atmosphere at the winter pole or sublime into the atmosphere from the summer pole.

As on Earth, much more sunlight falls on the equator of Mars compared with its poles, producing large temperature difference between these regions. These temperature differences vary according to season and distance from the Sun. Because the highly elliptical orbit of Mars carries it much farther away from the Sun during northern summer than in southern summer, heating during these seasons is quite different. As a result, the maximum temperatures in the atmosphere are found in the south polar region at Southern Hemisphere solstice, and the coolest temperatures are in the north polar winter nights. These differences have important effects on the Martian atmosphere and climate.

These latitudinal temperature differences heat the surface and atmosphere unevenly, generating a huge convection loop in the atmosphere, called a Hadley cell (see Figure 89). Hadley cell circulation develops as warm air expands and rises from the warm equatorial

Figure 89. Major circulation patterns within the Martian atmosphere. Arrows indicate direction of flow. Stationary eddies *(center view)*, or standing waves, are the result of *(A)* flow over topography and *(B)* temperature difference in different regions on the surface. (Modified from J. B. Pollack, Atmosphere of the terrestrial planets, 57–70 *in* J. K. Beatty, B. O'Leary, and A. Chaikin, *The New Solar System*, Cambridge, Massachusetts: Sky Publishing Corporation, 1981)

region, flowing at high levels in the atmosphere toward cooler polar regions. In the polar regions, the air cools and sinks to the surface. From the poles, it flows along the surface, back to warmer regions in the lower latitudes to be heated again. On Mars the ascending branch of the cell typically causes westward jets over the equator, and the descending branch of the cell generally give rise to a strong eastward jet of winds or polar vortex.

The characteristics of these cells show marked seasonal differences. Because of the high orbital eccentricity of Mars and the resulting large seasonal differences in heating, its Hadley circulation is stronger and wider during northern winter than during southern winter. Between early spring and fall, when the sun is over the mid-latitudes of Mars, is when Hadley cells develop in each hemisphere. Wind speeds of the polar vortex range from 90 to 100 m/sec (200 to 225 miles/hour) during these times. In Martian winter and summer, when the sun stays above the horizon at high latitudes, a single, huge Hadley cell develops in the warm hemisphere that extends across the equator. Typically in the winter hemisphere, the marked difference in temperature between the descending branch of the Hadley cell and the polar night causes strong eastward polar vortex winds to develop. Winds in this vortex can be nearly 160 m/sec (360 miles/hour).

As in Earth's atmosphere, this simple north-south circulation is broken into smaller cells by Coriolis forces. These sideways inertial forces are caused by the rotation of the planet. The equator of Mars is moving eastward at a much higher speed than the area around the poles. Therefore, the air masses moving away from the slow-moving poles appear to lag behind and shift westward, while pole-ward-moving air pulls ahead to the east. These deflections produce a spiral motion in the atmosphere that influences zonal flow (e.g., trade winds on Earth) and sets up turbulence and cellular patterns associated with cyclonic storms (Figure 90).

In the midlatitudes of Mars during winter and spring, strong temperature differences on the surface give rise to high- and low-pressure weather storm systems. These large-scale wavelike flow disturbances, called planetary waves, typically march eastward across the midlatitudes of Mars every 3 to 4 days. Viking Lander 2 detected the passage of these storms firsthand and discovered the timing to be remarkably regular. These weather systems produce sharp variations in wind speeds upward in the atmosphere, called baroclinic

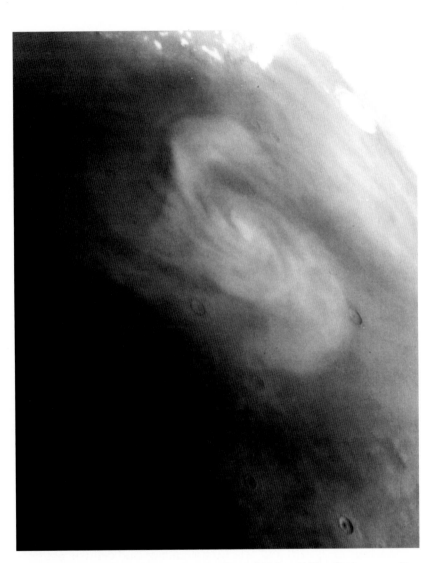

Figure 90. This late-summer storm is about 300 km (180 miles) across. It is located near the polar front of Mars, a strong thermal boundary that separates cold air over the pole from more temperate air to the south. This storm resembles storms seen in satellite pictures of extratropical cyclones near the polar front on Earth. Speeds of 20 m/sec (45 miles/hour) have been calculated for winds at the perimeter of the storm. (Courtesy NASA, Viking) (P-20805)

instabilities. On Mars these instabilities produce baroclinic eddies that help to mix the atmosphere by transporting heat poleward and vertically (see Figure 89). During times when the atmosphere clears, these storms are more frequent and intense. But during times of major dust storm activity this type of storm is less frequent and weaker. This suggests that the Martian atmosphere has two major dynamical regimes: one dominated by dustiness and one dominated by a relatively clear atmosphere.

Other weather systems of intermediate scale occur near the edges of the residual midsummer north polar cap (Figure 91). As Conway Leovy has suggested, "These systems, which produce familiar-looking moving comma-shaped cloud patterns, seem to be the dry counterpart of Earth's polar lows that form primarily over open water near the sea-ice edge during winter."

The extreme temperature differences (as much as 50°C [90°F]) between the day side and night side of Mars generates strong local winds called "atmospheric tides" or "thermal tides" (see Figure 89). These winds produce traveling waves in the atmosphere that follow the direction of the motion of the Sun in the Martian sky. Their amplitude typically depends on the amount of dust in the atmosphere. When the atmosphere is moderately dusty, proportionally more sunlight is absorbed on the daylight side, increasing its temperature. As a result, the difference in temperature between day and nighttime sides increases and the winds increase as well. But when the atmosphere becomes choked with dust, the mass of dust in the atmosphere is so great that it mimics the heat sink effects of Earth's oceans, decreasing the difference in daytime and night temperatures. When the temperature difference decreases, the velocity of these winds also declines.

Local wind systems are important on Mars because they sometimes can obtain a velocity high enough to move sand and raise dust. Local and regional winds are generated by differences in surface heating caused by differences in surface properties (e.g., surface reflectivity, the density of surface materials) from area to area. The largest of these areas, sometimes called thermal continents, are found in the low latitudes. They strongly influence longitudinal or regional circulation. In other places, heating on sun-facing slopes produces strong local winds. As the surfaces of these slopes heat during the day, the air above them expands and blows upslope. At night these slopes cool and the air becomes denser, causing it to sink and flow

downslope. The winds are strongest when Mars is closest to the Sun and are particularly important on large topographic features such as the giant volcanoes of Tharsis or in the Hellas impact basin.

Dust Storms and Ice Clouds

Astronomers have long observed clouds on Mars. Some of the clouds are white and some are the yellowish red color of Mars. The white clouds were thought to be water-ice clouds similar to those found on Earth. The reddish clouds were thought to be dust storms raised by howling winds. White clouds are found only in certain places during certain times and are very predictable, but the reddish clouds appear to occur sporadically.

Early astronomers found it difficult to predict when or where the dust storms would start or their eventual size and duration. They thought that there might be a correlation between dust storm occurrence and when Mars was closest to the Sun, during perihelion opposition. Perihelion opposition is when surface heating is at a maximum and near-surface winds are expected to be at the highest. However, early astronomers found it difficult to prove because during most of the year Mars is too far away from Earth to observe dust storm activity reliably. Without such observations a testable theory for dust storm formation is impossible to construct.

Since Mariner 9, atmospheric monitoring from spacecraft has begun to add long-term observations throughout the Martian year. These observations are beginning to provide the data needed to test the hypothesis of the early astronomers. As with the telescopic data,

Figure 91. *(Top)* This springtime dust storm originated on the edge of the north seasonal polar cap. These storms form as the seasonal polar cap sublimes, increasing the pressure and allowing more dust to be suspended and for a longer time, and increasing the temperature contrast between the cap and the surrounding ground; strong slope winds are generated off the high-standing layered deposits. The storm is moving as a front, outward from the central "jet" and extending about 900 km (540 miles) out from the north seasonal cap. *(Bottom)* For comparison with terrestrial dust storms, a picture taken from Earth orbit shows a similar dust storm extending about 1,800 km (1,100 miles) off the coast of northwestern Africa. The pictures are at the same scale. (Courtesy NASA/Jet Propulsion Laboratory/Malin Space Science Systems, Incorporated) (PIA02807)

these observations suggest that there is a general pattern to when and where dust storms occur. However, it is still impossible to predict exactly when and where individual storms will start. Most individual dust storms that develop are small and short-lived, though a few grow to engulf the entire planet. The smallest storms have winds that measure typically in the 14–32 m/sec (30–70 miles/hour) range and dissipate after several days. Many of these small dust storms occur around the edges of the polar caps and other places where temperature differences are high (Figure 91). Which of the small storms will develop into a global dust storm is still impossible to predict though recent observations continue to support the hypothesis of early astronomers that most develop at perihelion during southern summer.

When global dust storms start, they follow a predictable pattern of development over their lifetime. As they grow, they raise clouds of dust that absorb sunlight and warm the atmosphere. Heating produces temperature differences in the atmosphere that increase the magnitude of the daily swings in the atmospheric pressure and as a result generate strong winds. As the winds grow stronger, thick clouds of dust are raised and spread throughout the atmosphere. Because the dust in these clouds carries heat, their global spread eventually evens out day side and night side temperature differences (from 50 to 30°C [90 to 50°F]). As temperature differences decline, near-surface winds drop below the threshold velocity needed to raise dust to supply the dust storms. Consequently, the storms end.

Dust carried in the atmosphere plays a key role in determining what the temperature is in the polar regions and as a result strongly influences the water and carbon dioxide cycle on Mars. When and how much dust is suspended in the atmosphere is a controlling factor in whether ice is deposited on or removed from the poles of Mars. Besides carrying heat, dust acts as nucleation sites for condensation of frost. Without dust in the atmosphere, nucleation and growth of frost is very difficult. In addition, when dust clouds are thick, they block the warming effects of sunlight from heating the surface.

At present, during southern summer, when most of the global dust storms occur, the dust-laden atmosphere shields the southern cap from solar radiation, while in the north the abundant suspended dust provides nucleation sites for condensation and deposition of dusty frosts on the northern ice cap. During southern winter, dust storm activity is at a minimum, the air is clear, and the frost that condenses on the south polar cap contains little dust.

The dustiness of these ices influences their temperatures in the summer because it controls how much sunlight the ice will absorb. As a result, the bright clean ice in the south absorbs comparatively less sunlight and is cooler than the darker dusty ice in the north. These effects combine to produce maximum temperatures on the northern remnant ice cap that are about 45°C (80°F) higher than on the southern remnant cap. This is a remarkable difference considering that the North Pole is over 6 km (3.6 miles) higher than the South Pole and should be about 6°C (11°F) colder and that southern summer is at perihelion and gets 45 percent more sunlight. This difference in temperature may explain why there is a difference in composition of ices on these caps.

Mars Global Surveyor and the Mars spacecraft before it closely monitored dust storm behavior during their operation. Mars Global Surveyor observed five regional dust storms during its primary mission. These storms did not grow to engulf all of Mars but were large enough to have had a profound effect on the atmosphere's thermal structure. The largest of these storms raised the temperature in the atmosphere in the vicinity of the storm by 15°C (27°F). The two largest of these storms occurred during southern summer and grew to be more than 3,000 km (1,800 miles) across in the equatorial region. They were completely gone within 2 weeks.

Several global dust storms were observed firsthand from Mariner 9, the Viking orbiters, and Mars Global Surveyor, as well as the Viking lander spacecraft. Mariner 9 and Mars Global Surveyor observed the largest of these dust storms nearly 30 years apart. A global-scale dust storm was in full swing in 1971 when Mariner 9 began its orbital observation of Mars (see Figure 13). This storm obscured the surface for several months and provided atmospheric scientists with their first close-up view of a global dust storm on Mars. They were elated—the geologists were not because it obscured the surface. The atmosphere later cleared and all were happy. In June 2001 Mars Global Surveyor observed a global dust storm of about the same magnitude, but this time atmospheric scientists were able to watch as this storm developed from its beginnings. The storm started in the Hellas Basin region, where it alternately shrank and expanded for several days before exploding into a giant storm (Figure 92) that raised the temperature in the atmosphere by 30°C (54°F).

The Vikings observed three less-severe global dust storms over the 5 years of their operation. The orbiters tracked these storms from above (Figure 93) while the landers observed changes at the surface

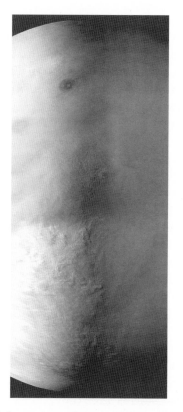

Figure 92. The global dust storm of 2001 as viewed by Mars Global Surveyor. This is the largest storm seen on Mars since that seen by Mariner 9 in 1971. It started as pulses of small storms in the Hellas Basin in June 2001 and quickly grew to enshroud the entire planet for several months. Noctis Labyrinthus can be seen in the center of the picture. (Courtesy NASA/Jet Propulsion Laboratory/Malin Space Science Systems, Incorporated)

as the storms passed (Figure 94). The landers measured wind gusts up to nearly 30 m/sec (65 miles/hour) and a marked darkening of the sky due to the suspended dust. Erosion of small piles of dust was observed at the lander sites, demonstrating that these storms could affect the surface.

Though dust storms are spectacular, condensation clouds and haze composed of water and carbon dioxide ice crystals are more common and widespread. This type of cloud has its own seasonal cycle that is as predictable as the dust seasonal cycle. Most Martian condensation clouds are similar to clouds found on Earth and form

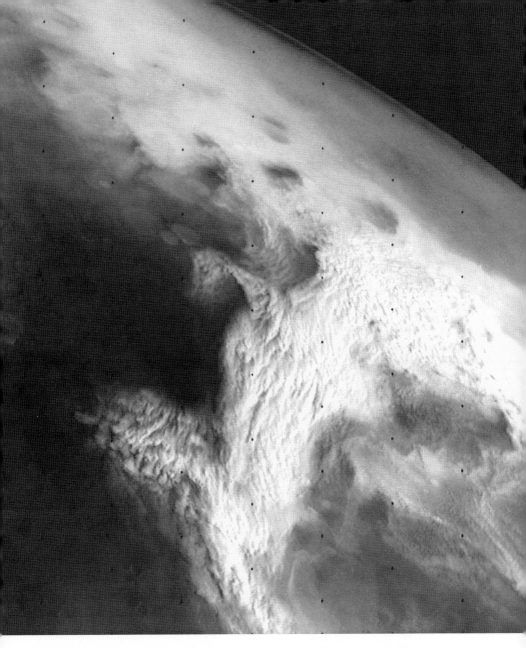

Figure 93. A dust storm over the Thaumasia region of Mars captured by Viking in 1978. This storm soon grew into the first global dust storm observed by the Viking orbiters. The storm occurred when Mars was at perihelion and heating of its surface was at a maximum. The picture shows an area about 1,400 km (840 miles) across. (Courtesy NASA) (PIA02985)

Figure 94. Sequence of pictures taken by Viking 1 Lander over three different days (Sol 1705, 1742, and 1853). The relatively low brightness on Sol 1742 is due to shading caused by a dust storm that passed over the landing site during that time. (Adapted from R. E. Arvidson, E. A. Guinness, H. J. Moore, J. Tillman, and S. D. Wall, Three Mars years: Viking Lander 1 imaging observations, *Science,* 22:463–468, 1983)

where warm moisture-laden air circulates to cooler regions of the atmosphere. As this occurs, the moisture in the warm air cools and condenses to form ice particles. Generally these ice particles nucleate around suspended dust grains. As a result, when condensation clouds form and precipitate their contents, they remove dust from the atmosphere. Consequently, the atmosphere over the winter pole, where frost is being deposited from the atmosphere, is generally dust free.

In northern summer (aphelion), low-altitude condensation clouds commonly form in a belt between 10°S and 30°N, the latitude where upward motions of the Hadley circulation are expected. This belt abruptly disappears as northern fall comes on and the atmospheric temperature and amount of dust it contains begin to increase. In addition, condensation clouds are commonly found as waves of clouds that form downwind of mountain ranges and high plateaus or above elevated regions at midday when surface heating is at its maximum (Figure 95).

Hazes or ground-hugging fogs are common in low areas on Mars (Figure 96). These develop during the coolest parts of the day (dusk and dawn) and typically vanish as these low areas warm (though some hazes are longer-lived and cover greater regions). In the fall, a

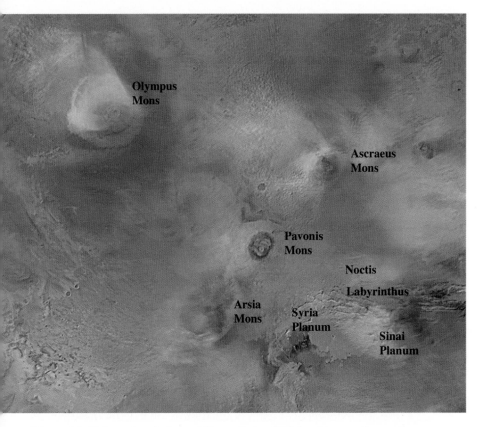

Figure 95. This regional view of Mars shows bluish white condensation clouds forming over elevated regions in the Tharsis region of Mars. The clouds appear when winds that carry warm, moist air from low regions blow up and over high regions. This cools the air and condenses the water vapor it contains into ice particles that form the clouds. (Courtesy NASA/Jet Propulsion Laboratory/Malin Space Science Systems, Incorporated) (PIA02653)

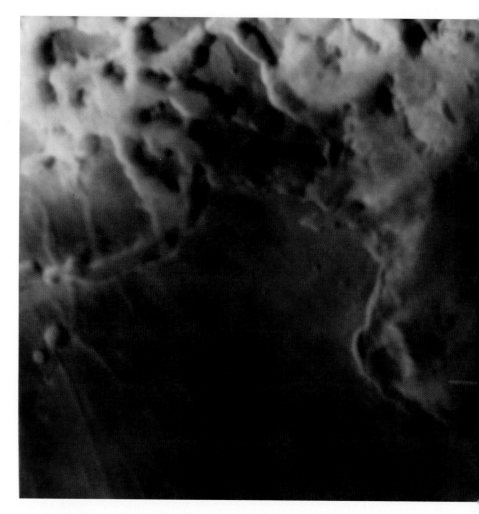

Figure 96. Early morning low clouds and ground fog are common in Noctis Labyrinthus. Such clouds are thought to be composed of water-ice crystals. (Courtesy NASA) (PIA03213)

haze or hood extends over each polar region, obscuring the surfaces (see Figure 76). This polar hood is more pronounced in the north than in the south and even periodically extends down to 35°N latitude.

As both types of clouds have been studied, scientists have begun to realize their importance to surface environment on Mars. These clouds strongly influence the thermal conditions in the lower atmosphere. In particular, suspended dust plays a critical role in the condensation and precipitation of frost. Thick dust clouds can shade the surface and increase the temperature of the atmosphere by absorbing solar energy.

Vertical Structure in the Atmosphere

The Martian atmosphere, like Earth's, shows systematic vertical variations in its structure. The terminology used to discuss the general structure of planetary atmospheres at different levels is derived from Earth's atmosphere and is based primarily on temperature effects. However, these temperature effects are caused by different mechanisms in different atmospheres. According to Richard Zurek, a Mars atmosphere expert at Jet Propulsion Laboratory, "On Earth, the atmospheric regions are distinguished by vertical variation of temperature due largely to the presence of ozone. On Mars, the comparable absorber is airborne dust, but how much and where it occurs in the atmosphere can change enormously and unpredictably. The height of any dust haze on Mars will vary seasonally, regionally and episodically." Because the ozone content in Earth's atmosphere stays relatively constant, its atmosphere is more structured and stable than that of Mars; the tremendous variability of dust in the Martian atmosphere over the course of each year causes its thermal structure to vary considerably. When the lower atmosphere of Mars is dust free, its temperature profile resembles that of Earth's troposphere where temperatures continuously decrease upward. When the Martian atmosphere is dusty, it may have a more nearly uniform temperature profile and resembles Earth's stratosphere.

The Lower Atmosphere

"The lower atmosphere is defined as the region from the ground to about 45 km [27 miles]. This region is influenced strongly in its lower reaches by radiative heat exchange with the ground; it is also the region most strongly heated by airborne dust. Most of the available temperature observations are of this region, sometimes called the Martian troposphere," explained Richard Zurek. Most of the dynamic weather phenomena are observed in the lower 6–10 km (3.6–6 miles) of this part of the atmosphere because of such heating; at higher altitudes the air is thinner and more transparent to sunlight and little energy is absorbed in each kilometer of height. Daytime surface heating generates convection in this lower 6–10 km (3.6–6 miles) of the atmosphere. During the night the convection collapses and strong temperature inversion develops in the lower kilometer. Temperature changes in the lower atmosphere send strong time-dependent oscillations in temperature that are propagated upward in the atmosphere.

In spite of the decreased pressure upward, winds, wispy clouds, and hazes are observed at altitudes a few tens of kilometers above the surface (Figure 97). At these altitudes the temperature is typically low enough for carbon dioxide ice, as well as water ice, to be stable and to form clouds (Figure 98).

Though temperature and pressure generally decrease continuously upward through the lower atmosphere, temperature inversions have been observed above the nightly inversions in the lower few kilometers. These inversions occur mainly when the atmosphere is at its dustiest and absorption of solar energy is at its maximum or in southern summer when the strength of Hadley circulation is greatest and strong low-altitude winds carry cold air from the North Pole southward. Such inversions can affect the stability of the carbon dioxide ice and result in the formation of condensation clouds (Figure 99).

The Middle Atmosphere

Above the lower region in the Martian atmosphere, where turbulent winds and cloud-forming activity are common, is the middle atmosphere, a region of nearly uniform temperature and smooth airflow, lacking turbulence. Richard Zurek indicated that "the middle atmosphere, defined as the region extending from the lower atmosphere to the base of the thermosphere (45 km to 110 km [27 to 66 miles]). This region is sometimes called the Martian mesosphere."

From the top of the Martian lower atmosphere up to about 110 km (66 miles) the temperature remains nearly constant. On Earth the temperature in the mesosphere rises with increasing altitude due to absorption of ultraviolet light by ozone; in the Martian mesosphere, with little ozone in its oxygen-poor atmosphere, there is limited heating by this mechanism. In spite of a lack of a temperature differential to drive winds in this region, winds still occur but are generated by atmospheric waves from surface topography and thermal tides. Thin, wispy layers of cloud have been observed as high as 110 km (66 miles).

The Upper Atmosphere

From about 110 to 200 km (66 to 120 miles) altitude, the temperature rises rapidly. This region of the atmosphere is called the ther-

Figure 97. Pink stratus clouds as seen from the Mars Pathfinder Lander. The clouds are at an approximate height of 16 km (10 miles) and are moving at about 7 km/sec (15 miles/hour). The clouds consist of water ice condensed on reddish dust particles suspended in the atmosphere. Mars Pathfinder took this picture about 40 minutes before sunrise. (Courtesy NASA) (PIA00784)

Figure 98. An oblique view across Argyre Planitia toward the horizon about 19,000 km (12,000 miles) away taken by Viking 1 Orbiter. The brightness on the horizon is mainly due to a haze composed of crystals of carbon dioxide that condensed at 25 to 40 km (15 to 24 miles) altitude. (Courtesy NASA) (P-17022)

mosphere and corresponds to similar regions in the atmospheres of Earth and Venus. The rapid rise in temperature in this region is caused by a combination of the heat produced by the absorption of ultraviolet light and the low efficiency of the gases at that altitude in radiating their heat away. At such high altitudes the temperature of the atmosphere, a measure of the average speed of randomly moving gas molecules, becomes an abstract concept. It loses its everyday meaning, because the air is so thin that it contains little heat.

Above about 110 km (66 miles) is a region of the thermosphere called the ionosphere. In this region of the atmosphere, ultraviolet sunlight is so intense that it strips electrons from atoms and molecules to create an ionized gas layer. These electrons carry a negative electrical charge; in Earth's atmosphere they can significantly affect some radio waves. Though carbon dioxide is the main ionized component of the Martian ionosphere, oxygen ions are actually the most abundant.

Aside from its electrical effects, the ionosphere makes its presence felt in another way: it glows. This airglow, as it is called, is caused by the same electrochemical reaction that causes ionization. Instead of merely stripping an electron from a molecule, ultraviolet light may change the energy state of the molecule, causing it to release some of its energy as light. The intensity of these emissions can vary considerably and is a direct result of changes in solar radiation. In 1972 Mariner 9 observed a variation of 20 percent in magnitude of emission from atomic hydrogen in the Martian ionosphere. In addition, independent of intensity level, each type of molecule emits a characteristic light wavelength. This property is used as a remote sensing tool to detect the presence of specific components.

The lower part of the thermosphere is a thin, homogeneous assemblage of gases mixed by winds generated from daily and seasonal temperature variations. In the calm upper reaches of the Martian thermosphere, above about 125 km (75 miles), where neither winds nor convection mixes the various elements, these components stratify. The heavier ones, such as molecular nitrogen, carbon dioxide, and oxygen, settle to form layers below those of lighter atomic gases, such as hydrogen and light isotopes of other gases.

Above the thermosphere is the furthest province of the atmosphere—the exosphere, a bleak environment intensely irradiated by the sun. The density of atmospheric components in this region of the atmosphere is so low that individual particles may collide only oc-

Figure 99. These clouds over the crater Kepler are thought to have formed where an inversion layer formed when the Kepler-related disturbance allowed a portion of the warmer lower layer of air to break up through cooler altitudes in the atmosphere. This resulted in condensation of water in the warm air and formation of a plume about 150 km (93 miles) long. Inversion layers are formed where cooler, denser air is trapped beneath a layer of warm air. The picture was taken during midwinter in the Southern Hemisphere by Viking 1. (Courtesy NASA, Viking)

casionally: molecules near the surface of Mars travel less than a centimeter before they collide, but the gases in the exosphere may travel 5–10 km (3–6 miles) before they hit another particle. With few collisions to slow them and heated by sunlight, these light gases move several times faster than molecules in the lower atmosphere. As a result they are sprayed up toward space. Gravity eventually pulls most of the speeding atoms back to Mars, though a fraction moving at a speed in excess of the escape velocity of Mars breaks away into interplanetary space. This process is responsible for significant changes in the composition of the Martian atmosphere and is discussed further in the section on origin and evolution of the atmosphere later in this chapter.

Just where the exosphere ends and interplanetary space begins cannot be defined with certainty; it is merely a matter of degree. Molecules and atoms become more and more scarce until, somewhere about 200 km (120 miles) above the surface, there are none. Beyond the exosphere of Mars lies the void of space.

This is not the case on Earth. The exosphere is surrounded by a far larger aura, the magnetosphere. The magnetosphere is not, strictly speaking, part of the atmosphere, but it is bound up with its working and with life on Earth. The magnetosphere is produced because of Earth's magnetic field. The invisible lines of force created by Earth's magnetic field either trap high-energy particles from the solar wind (the charged particles that speed outward from the sun at 1½ million km/hr [900,000 miles/hour]) or deflect them at the edge of the magnetosphere. As a result, Earth has a protective line of defense from the Sun's deadly radiation effects. Mars currently has no global magnetic field and hence no magnetosphere. But Mars may have had such a protective cover early in its history that disappeared as its core solidified.

Composition

The current Martian atmosphere is composed mainly of carbon dioxide—about 95.3 percent. Lesser amounts of nitrogen (2.7 percent) and argon (1.6 percent) are also present, along with trace amounts of diatomic oxygen, ozone, carbon monoxide, water, heavy water, nitric oxide, neon, krypton, and xenon. These components are not uniformly distributed throughout but systematically change with altitude and location. There are also signs in the relative abundance of these components that the composition of the Martian atmosphere has changed with time.

Carbon Dioxide

Carbon dioxide, the main component of the Martian atmosphere, is readily photodissociated (broken apart by sunlight), producing carbon monoxide and atomic oxygen. But there are only small amounts of carbon monoxide in the Martian atmosphere. To maintain the current level of carbon monoxide, photodissociation is counterbalanced by a series of reactions between carbon monoxide and atomic oxygen that reform carbon dioxide as rapidly as it is destroyed. These reactions are greatly hastened by the catalytic action of "odd" hydrogen (odd molecules are groups of atoms that stay together and take part in chemical reactions as if they were a single element), a product of the photodissociation of water vapor.

Water and Its Products

At any one time, the Martian atmosphere contains only a tiny amount of water vapor, collectively only about enough to fill Henry David Thoreau's famed Walden Pond. Like carbon dioxide, this water vapor is in a constant state of flux, continuously being formed and destroyed. Its abundance is modulated by the rate at which it exchanges with surface reservoirs, in particular the polar ice caps. The ice caps provide a summertime source and wintertime sink for water. The polar caps are probably not the only source of water vapor. Most likely there is also an enormous amount of water trapped in the soils (regolith) and rocks, especially in the polar regions. Under current climate conditions, exchange with this reservoir is limited. However, during periods of milder climate, this reservoir may release huge amounts of water and other volatile components into the atmosphere.

Odd hydrogen is one of the most important products of the dissociation of water. It is important not only because of its role in the catalytic reactions that maintain the carbon dioxide levels in the atmosphere (mentioned earlier) but also because of it role in producing oxidants such as hydrogen peroxide and other polyoxides. Though these oxidants are produced by reactions in the atmosphere, they are important to the chemistry of the Martian regolith. Remarkably, hydrogen peroxide is among the most likely molecules to condense onto the Martian surface. It condenses nearly as readily as water vapor does from the Martian atmosphere. But its life on the surface is short. Sunlight easily breaks it down in a few hours. At night this process reverses and hydrogen peroxide quickly reforms. While on the surface, hydrogen peroxide reacts with the materials in the regolith to produce other polyoxides. It was these compounds that may have been responsible for the gases detected by the Viking labeled release experiment. Some of these oxidants readily release oxygen when combined with water, such as that contained in the "chicken soup" used in this experiment (described in chapter 7). In addition, some of these oxidants are a death sentence for organic molecules. Reactions between organic molecules and these polyoxides rapidly destroy the organic molecules. Consequently, the relationship between water vapor, odd hydrogen, and the production of oxidants combines to make the surface of Mars a very hostile environment for life.

Hydrogen

Molecular hydrogen is also found in the Martian atmosphere, a product mainly of the dissociation of water vapor. Tobias Owen at the University of Hawaii found that abundance of the heavier isotope of hydrogen is higher by a factor of 5–6 than predictions from solar nebula studies. Most likely this is the result of the greater loss of the lighter isotope from the atmosphere. As with other light gases, hydrogen escapes from the top of the Martian atmosphere. Because of its low density, it has the highest escape rate of any gas and its lighter isotope has the highest escape rate of all. Considering the current loss rate and the current inventory of hydrogen, Owen estimated that to produce the observed 5 times enrichment "over 99 percent of the water available to the Martian surface must have been destroyed with subsequent hydrogen escape." Based on this loss rate the total inventory of Martian water is calculated to be an equivalent of a global ocean 3 m (10 feet) deep. This is much smaller than predicted from other measurements, as is discussed later.

Because the prediction of such a small inventory of water is inconsistent with most other evidence, something is probably wrong with the calculation. Owen suggested that these calculations underestimate the initial inventory of hydrogen because they assume that the loss rate has stayed the same over the history of Mars. He has offered a solution, suggesting that the Martian environment was different in the past in a way that produced a higher loss rate of hydrogen. He pointed out that there are three possible mechanisms that could produce this high loss rate: "oxidation of the crust, enhanced solar ultraviolet radiation, and a general warming of the atmosphere." Owen favors atmospheric warming caused by climate change because of the abundant evidence that the Martian climate was milder in the past. Milder conditions would increase the loss of components from the top of the atmosphere because their loss rates are dependent on temperature.

Oxygen

Oxygen tells its own story about the inventory of water on Mars, as well as the evolution of its atmosphere. As with carbon, the relative abundance of light isotopes of oxygen is the same as those found in other solar system bodies. In theory, the light isotopes of all gases that can escape from the top of the atmosphere should have been selectively lost from Mars at a much greater rate than the heavier isotopes. Most scientists think that the light isotopes of oxygen and carbon were also lost, but their loss was masked by replenishment from a large reservoir. They suggest that this reservoir of carbon dioxide and water was trapped in the regolith, and under the right conditions it could exchange freely with the atmosphere. It has been calculated that to maintain the current level of oxygen, the Martian regolith must have released enough water for an equivalent of a global water ocean 13 m (43 feet) deep.

Small amounts of ozone are found in the Martian atmosphere. Most oxygen in the Martian atmosphere is tied up in carbon dioxide and water vapor. Although these compounds are easily photodissociated and quickly recombine, some atomic oxygen produced by photodissociation escapes to form ozone. This ozone is typically short-lived in the Martian atmosphere. It is rapidly destroyed by water vapor through reactions with odd hydrogen. For this reason, ozone is only found in the atmosphere where the humidity is very low, such as at the winter pole. In places where water vapor is abundant, such as at the summer pole, ozone quickly disappears from the atmosphere.

Nitrogen

Though nitrogen is the dominant gas in Earth's atmosphere, it is only a minor component of the Martian atmosphere. Nitrogen is important in the chemistry of both atmospheres, especially in the ionosphere. On Mars nitrogen compounds break down or dissociate readily in the upper atmosphere to produce odd nitrogen that reacts with oxygen, the dominant ionized component in the ionosphere.

Nitrogen is important for another reason. It provides clues to the evolution of the atmosphere. The heavy isotopes of nitrogen are twice as abundant in the Martian atmosphere as those in the atmosphere of other terrestrial planets. This anomaly is thought to be due to the selective escape of the light isotopes of nitrogen compared with the heavy ones from the top of the Martian atmosphere. Considering the current amount of nitrogen and its estimated escape rate, it has been calculated that nearly 90 percent of Martian nitrogen must have been lost over its history. This requires an initial equivalent of about 1.3 mbars (0.02 pounds/square inch) of nitrogen.

This estimate provides an additional clue to the evolution of the atmosphere. Because the carbon to nitrogen ratio for Mars, Earth, and Venus should all be about the same (they all formed from the same solar nebula), the abundance of nitrogen on these planets can be used to estimate the original inventory of carbon dioxide. Using the predicted initial 1.3 mbars (0.02 pounds/square inch) of nitrogen just mentioned, calculations of the initial pressure of carbon dioxide suggest that there may have been as much as 78 mbars (1.2 pounds/square inch) of carbon dioxide in the early atmosphere.

Noble Gases

As do many other gases in the Martian atmosphere, the relative abundance of noble gases (i.e., inert gases such as argon, xenon, and krypton) and their isotopes shows anomalies that suggest that the Martian atmosphere has undergone substantial changes over its history. For example, some of the isotopes of the noble gases argon, xenon, and krypton are radiogenic and are the decay products of radioactive potassium and iodine. Other isotopes of argon, xenon, and krypton are nonradiogenic and are thought to be part of the original inventory of Martian volatiles. The relative abundance of the radiogenic isotopes is anomalously high compared with their nonradiogenic cousins. This anomaly is thought to reflect the effects of processes that operated early in Mars history (such as impact and solar wind erosion) that blew away the Martian atmosphere, hence reducing the total inventory of gases trapped by Mars. In the case of noble gases, the new atmosphere was rebuilt with nonradiogenic isotopes in the reduced inventory of volatiles trapped inside Mars and radiogenic isotopes produced by elements (radioactive potassium and iodine contained in the rocks) that had not been reduced by the loss of the old atmosphere. As a result of the addition of these radiogenic noble gases, currently the fraction of these gases is anomalously high compared with the relative abundance of other gases in the atmosphere.

Origin and Evolution of the Atmosphere

Mars has had some form of an atmosphere for most of its history, but exactly where it came from and how and when it initially formed is not completely known. It is clear that the origin of the Martian atmosphere must be closely tied to the formation of the solar system.

Central to the question of the origin of the atmosphere is whether Mars accreted while the nebula was still present or after the nebula had been blown away by the intense solar wind streaming from our young Sun. If the nebula was present, the current atmosphere would be mainly composed of gas captured directly from the nebula and trapped in the planet as it grew. If the nebula had been blown away, atmospheric gases could have been added as a thin volatile-rich veneer on Mars by late-stage accretion of comets and volatile-rich asteroids, or from gases already incorporated in accretionary materials before they formed Mars. Gas-carrying materials could have picked up their volatiles either by their being implanted by the early intense solar wind or by sorption of gases from the nebula before it was swept away.

Most scientists favor the hypothesis that Mars accreted in the gas-rich environment of the nebula. They suggest that it follows that the Martian atmosphere is composed of gases that were trapped before the nebula was cleared by an early intense solar wind or blasted away from the planet by the collision of asteroid-size debris (with primordial Mars). The gases trapped inside Mars as it continued to develop began immediately to leak out to form the atmosphere.

During that time, Mars began to belch huge amounts of hydrogen, a product of reactions between its primordial water and metallic iron. This enormous flood of hydrogen swept other volatile components along with it. Based on estimates of the amount of hydrogen produced during oxidation of the mantle, enough hydrogen was liberated from inside to produce (if combined with oxygen as water) a global ocean 100 to 200 km (60 to 120 miles) deep. This hydrogen would have stayed with Mars for only a very brief time. It would have escaped rapidly into space, boiling out of the steaming hot primordial atmosphere, or blown away by the intense ultraviolet radiation from the young Sun, or blasted off by the occasional giant impact event. Hydrogen-producing reactions in the interior quickly ran their course, and the rate of hydrogen degassing soon fell rapidly.

The rate of degassing of most other components from a planet's interior is strongly influenced by its interior temperature. Mars was still hot as degassing of hydrogen declined. As a result, volatiles continued to leak out of Mars at a relatively high rate. For a short time, dramatic thermal events, such as the formation of the core and differentiation, kept the interior hot and the rate of degassing high. It has been calculated that during that time enough gas was lost from the interior to produce an atmosphere 100 times the mass of the current one.

Much of the carbon dioxide in the Martian atmosphere is the result of gases exuded during volcanic eruptions throughout the planet's history. None of the products of the early outgassing of Mars is thought to have stayed with the planet. These gases emerged during the first half billion years of Martian history when Mars was undergoing intense bombardment by debris left over from the final stages of accretion. The enormous blasts from the ancient impact events, especially the largest ones, would have "blown off" any early atmosphere. Only one or two basin-forming impact events would easily have been enough to remove the entire primordial atmosphere.

A Changing Climate

By about 4.0–3.8 billion years ago, Mars entered an era when its atmosphere and surface stabilized. For the rest of its history, the rate of addition or subtraction of gas to the atmosphere of Mars has been much less dramatic than during its first half billion years. That was a time when heavy bombardment by large asteroids and comets came to a close and the repeated blowing off of the atmosphere ceased, although some estimates suggest that impact erosion of the atmosphere was still responsible for the loss of 50–90 percent of the volatiles that outgassed from the planet for the next several hundred million years of its history. During that time the interior was cooling rapidly, causing the rate of degassing to fall dramatically, although there is evidence that the interior still continued to provide large amounts of gas to the atmosphere through the extensive volcanic activity during that early period. The processes that slowly removed gases from the atmosphere were also in operation (solar wind erosion and photochemical processes). Three and a half billion years ago, these processes had helped to change the early dense atmosphere to one that is more like what we see today. Evidence for this early dense atmosphere is preserved on Mars in the composition of its atmosphere and the effects it had on ancient landforms.

The Early Earth-Like Climate: Martian Garden of Eden
The surface of Mars records evidence that about 3.9–3.8 billion years ago the Martian climate was very different than it is today, though how different is the subject of considerable debate. Fraser Fanale at the University of Hawaii, a leading researcher in Martian climate studies, has summarized the most convincing lines of this type of evidence. He listed as evidence "(1) the difficulty of forming valley networks by any agent other than running water; (2) the almost complete restriction of valley networks to the ancient cratered highlands; (3) the requirement that shallow regolith temperatures be close to 273°K (freezing point of water) for valleys to form irrespective of origin by sapping or precipitation; and (4) evidence of enhanced obliteration of features early in the planet's history." Fanale pointed out that one the most important lines of evidence, the formation of valley networks, requires only a groundwater system for their formation. He explained, "the heat flow expected for early Mars would cause groundwater to be closer to the surface than present" though surface temperatures were subfreezing. He added that groundwater could easily have reached the surface without freezing during the time the early dense atmosphere was present because "the surface temperatures may have been higher because of the thick CO_2 atmosphere." If this was the case, then the intense erosion during that early period required only a substantially thicker, though not necessarily warmer or wetter atmosphere than the current one.

Bruce Jakosky and Roger Phillips of Washington University in St. Louis think that the climate initially may have been relatively warm and wet and then shifted at the end of that early episode to colder, drier conditions. They point to the "initiation of U-shaped valley forms at the downstream ends of V-shaped valleys, suggesting an evolution of valley-forming erosional mechanisms from water to ice-related. This period of U-shaped valley formation, and valley network in general ended rapidly" at about 3.5 billion years ago. U-shaped valleys are diagnostic of being cut by ice and are a common feature in areas of alpine glaciation on Earth.

Jakosky and Phillips pointed out that there also may be a connection between this early mild climate and the formation of Tharsis. This connection may explain why mild conditions persisted for several hundred million years after heavy bombardment. Recently collected evidence suggests that most of Tharsis was built early in Martian history and was probably contributing its gases to the Martian environment even before the end of heavy bombardment. The patterns of many of the ancient valley networks in the Tharsis region are orientated radial to Tharsis, a result of slopes created by downwarping of the lithosphere from the weight of this massive pile of volcanic materials. Most of the large-scale geological features related to surface water are concentrated in this trough. Because the magma that built Tharsis also contained substantial quantities of volatiles, the early atmosphere would have been replenished repeatedly until the early intense era of volcanism that built Tharsis declined. Jakosky and Phillips speculated about the role of Tharsis in this early period of mild climate and suggested that "the timing of the valley networks formation may be more than coincidental. Tharsis volcanism may have supplied gases that helped maintained the climate conducive to weathering and erosion, and the cessation of volcanism may have allowed other processes to begin the removal of much of the atmosphere."

The surface is not the only source of evidence about this early dense atmosphere. Additional evidence for an early dense atmosphere is provided by the relative abundance of components found in the Martian atmosphere (see discussion in the previous section). Many key components of the current Martian atmosphere show significant differences between their relative abundance and the relative abundance of the same components in the solar nebula (i.e., gas trapped in Mars from the nebula) determined from analysis of the composition of the Sun and of ancient meteorites. These anomalies have been attributed to physical and chemical processes that selectively subtract or add individual components. Calculations based on comparisons between the relative abundance of components (in the Martian atmosphere and the Sun) suggest that the density of the early Martian atmosphere may have been as much as 100 times greater than the current atmosphere. For example, current oxygen isotopic abundance requires that the early dense Martian atmosphere had 10 times more carbon dioxide than at present, and estimates of the total amount of interior degassing predict 100 times more carbon dioxide than is in the current Martian atmosphere. These estimates predict an early Martian atmosphere about 5 times denser than Earth's current atmosphere. What are the effects of such a dense carbon dioxide–rich atmosphere on the Martian environment?

One effect is heating. The increased density of a carbon dioxide–rich atmosphere should have had a warming effect on the surface environment of Mars through the process of greenhouse heating. Greenhouse heating occurs because carbon dioxide is transparent to sunlight but absorbs thermal infrared energy (heat). As a result, sunlight easily penetrates the Martian atmosphere to warm the surface. The warm surface reradiates this heat in the form of thermal infrared energy. The carbon dioxide–rich atmosphere readily absorbs this thermal infrared energy instead of allowing it to pass through. This heats the atmosphere. Based on estimates of the density of the early thick atmosphere, it could have been heated to a temperature that was warm enough to sustain liquid water on the Martian surface.

Cooling is another effect. There are other competing processes that would counterbalance greenhouse heating: one a direct consequence of the dense atmosphere and the other a result of the behavior of the early Sun. Because the early atmosphere was dense, it also should have been cloudy. Clouds prevented solar heating of the surface by shading it and reflecting sunlight back into space. But whether there were enough clouds in the Martian skies 3.8 billion years ago to counterbalance greenhouse heating is still a matter of debate. In addition to the effects of increased cloudiness, most stars, like the Sun, go through an early period when they radiate less sunlight. By some estimates, during the time the early dense atmosphere was present on Mars, the Sun was radiating only about 90 percent of its current level of sunlight. As it has on all the other planets, this would have had significant effects on the surface temperature on Mars. It has been calculated that such a reduction in sunlight would have been enough to cause the freezing of Earth's oceans.

Did any one of these processes win this tug-of-war of heating and cooling of the surface? Some scientists suggest that the effects of greenhouse heating in the early dense atmosphere combined with heat flow from the interior outweighed the effects of decreased solar luminosity and cloudiness. They suggest that the atmosphere was warm, wet, and dense, resulting in a mild, Earth-like climate. At odds to this view, other scientists suggest that the evidence for mild conditions during this early climatic epoch has been overinterpreted. They hold that the available evidence indicates only that the early atmosphere was thicker but not significantly warmer or wetter than the current one.

It is clear that the early atmosphere that produced the extensive erosion in the ancient terrain and left chemical anomalies was very different than the current atmosphere. If it really was much denser, why did the density plummet to its current level? Scientists studying Mars have found that there are processes that remove gas from the top of the atmosphere and others that trap gases on the surface and in the regolith. Gases lost from the top of the atmosphere are lost forever, but under the right conditions gases trapped on the surface and in the regolith can rejoin the atmosphere. The rate at which most of these processes operate is temperature dependent. As a result, if the early atmosphere was warm, its loss would have been substantially accelerated. When higher temperature is factored into the calculation for the loss of gases from the early Martian atmosphere, its effects easily account for the reduction in density to the current state.

Loss of gas from the top of the atmosphere is generally through the effects of either sunlight or the solar wind. Of these loss mechanisms, the effects of sunlight are the most important in causing permanent changes to the Martian atmosphere. Loss occurs when gas molecules, such as carbon dioxide, at the top of the Martian atmosphere absorb enough solar energy to excite their atoms to break apart at the escape velocity of Mars, or when atoms absorb enough sunlight to be heated and accelerated to escape velocity. In the rarified upper region of the atmosphere, when one of these high-velocity particles is aimed upward and its pathway is free of other gas atoms, it may shoot out of the atmosphere and be permanently lost to space. The solar wind also removes gas from the top of the atmosphere, though at a lower rate. The solar wind collides directly with gas at the top of the Martian atmosphere. Without a magnetic field for protection, some of these collisions are so powerful they can knock gas particles out of the grasp of Martian gravity. However, early in its history when Mars had a substantial magnetic field, loss by the mechanism would have been negligible. This is consistent with the pristine proportions of nitrogen and argon isotopes trapped in Martian meteorite ALH84001, which indicate that solar wind stripping in the atmosphere had not yet occurred when these gases were trapped in this ancient rock.

On massive planets, such as Earth and Venus, these processes are relatively unimportant because their strong gravity fields firmly bind atmospheric gases to these bodies. In part, this is why Earth has retained much of its water. On smaller planets, such as Mars, these processes have substantial effects on the loss of their atmospheres. For Mars, its relatively weak gravity field allows gases from its atmosphere to escape at a rate that, over time, has changed its atmospheric composition. These loss processes favor the selective escape of the lightest gases because less energy is required to accelerate them to escape velocities and loss of gases occurs in a part of the atmosphere where light gases are concentrated. As a consequence, these processes fractionate the Martian atmosphere and deplete it of its lightest components (e.g., the light isotopes of light gasses such as hydrogen).

Not all loss processes involve permanent escape and loss of gases from Mars. Some processes remove gases from the atmosphere, but do not eliminate them from the planet. Most of these lock up gases permanently in rocks, but others store gases temporarily on the surface or in the regolith until surface conditions change. Under the right conditions volatiles derived from the atmosphere can be incorporated into the structure of mineral grains or locked in rocks such as carbonates. These gases are permanently lost to the atmosphere. Other volatiles are trapped in regolith in subsurface cracks and on grain boundaries, the equivalent of a global layer or ocean 600 to 1,400 m (1,970 to 4,590 feet) deep. Volatiles trapped in these places can readily enter or exit the regolith during times when the climate is mild and infiltration or escape is easiest. But during times of cold climate conditions, such as the present, volatiles do not easily enter or escape from the subsurface.

In summary, after the stabilization of the surface and atmosphere about 3.8 billion years ago, Mars developed a relatively dense carbon dioxide–rich atmosphere. This dense atmosphere caused mild climatic conditions that produced a surface environment much more dynamic than that currently on Mars. Within a few hundred million years, a substantial portion of the early dense atmosphere was lost. Although a considerable portion escaped from Mars, a substantial portion was also trapped on the surface and in the regolith; it is temporarily lost, but most is retrievable. This early dense atmosphere hastened its own demise, increasing its own temperatures and as a result also increasing the loss rate of its gases. Loss of these gases quickly changed the climate, producing colder and perhaps dryer conditions. In the next section, we will explore evidence that after the loss of this early dense atmosphere and mild climate, the climate has oscillated periodically between current conditions and milder conditions.

Quasiperiodic Climate Changes
There is evidence that, after the loss of the early dense atmosphere, the Martian climate periodically fluctuated between harsh, cold, dry times and milder periods. The thinly layered polar deposits are generally regarded as the single most convincing piece of evidence for periodic climate changes on Mars (see chapter 4 for a detailed description of these deposits). These deposits are thought to be composed of layers of dusty ice whose individual layers are distinguished by the relative amounts of dust and ice each contains. The characteristics of each layer are influenced by the effects produced by cyclic changes in the motion of Mars.

Besides the rhythmic nature of their layering, the polar layered deposits have another notable characteristic: they appear to be no more than a few tens or hundreds of millions of years old. Does this mean that the conditions that produced these deposits have existed only for the past several hundred million years of Martian history? Probably not. Instead it may mean that conditions that created these deposits have been episodic. Most scientists expect that the geologic axiom "the present is the key to the past" holds equally well for Mars as it does for Earth and that conditions that produced these deposits have existed since the formation of the current atmosphere. If this is the case, evidence of older layered deposits should be preserved somewhere on Mars. Indeed, scientists think they have found remnants of older layered deposits that resemble the polar layered deposits. The highly eroded, southern etched and pitted terrain is regarded as an example of this type of deposit. These deposits are well over 3 billion years old. What would the climate have been like to produce these deposits?

A debris mantle, thought to be deposits of windblown materials, surrounds the northern polar layered deposits down to the middle latitudes. This mantle, originally described by Laurence Soderblom and his colleagues at the U.S. Geological Survey, appears to be uniformly draped over materials equatorward of the layered terrain. In many places the layered deposits have been deposited on top of debris mantle deposits.

Other layered deposits are also found outside the polar regions on Mars. Michael Malin and Ken Edgett have extensively studied intermediate-age layered deposits they have found in western Arabia Terra, the intercratered plains of northern Terra Meridiani, and Valles Marineris. They concluded that these deposits formed when the climate of "Mars was very dynamic and may have been a lot more like Earth than many of us had been thinking." They suggested that "The nearly identically thick layers would be almost impossible to create without water," although other scientists consider the wind to be an equally likely candidate that would have required a less-dramatic climate change. On Earth such uniformly thick, rhythmic layering typically indicates an environment where sediments transported into a region, whether by water or wind, can settle out of the medium that carried them to the deposition site. Whether the Martian layered deposits are composed of dust deposited from air or courser sediment deposited in lakes or seas is important. Each requires different environmental conditions. For these deposits to be lake or sea sediments, the surface environment would have had to be warmer than the present, but for these deposits to have formed like the polar layered terrain, all that was needed was for the atmosphere to be denser. A closer look at these deposits is required to determine which scenario is the correct one, though either way it is clear that these deposits formed at a time when climate conditions were very different than they are today.

What caused these periodic climate fluctuations? Most scientists think that climate changes on Mars are caused by quasiperiodic changes in the orbital motion and axial orientation of Mars (mainly oscillations in obliquity, orbital eccentricity, and precession of the equinoxes). Orbit and axial characteristics strongly influence how much and where sunlight falls on Mars and, as a consequence, temperature distribution over the surface. As with the other planets, cyclic changes in the orbit and rotation axis orientation of Mars are produced by the gravitational pull from other bodies in the solar system. Of the astronomical variations produced by this cosmic tug-of-war between the planets, obliquity has the greatest influence on the climate of Mars. Currently, the obliquity of its orbit oscillates with a period of 120,000 years about its current mean. Its amplitude also varies, but with a period of about 1.3 million years. Other astronomical variables also have an effect on climate, though much less pronounced. Their main effect is to either slightly amplify or slightly diminish the effects of obliquity.

Obliquity is important because it controls the orientation of the poles of Mars and, combined with the distance to the Sun at perihelion, strongly influences how much and where sunlight falls on its surface and the timing and intensity of its dust storms. In particular, the characteristics of these motions control the temperature of the polar regions, which is important because that is where a huge inventory of volatiles is stored. Surface temperatures in these regions are always near the frost point of carbon dioxide and water. As a result small temperature changes can have enormous consequences, causing either huge amounts of the atmosphere to freeze or frost to sublime into the atmosphere. To a certain extent, this happens every year on Mars and is responsible for condensation flow in the atmosphere.

What effect do changes in obliquity have over the longer term? At times of high obliquity, summers of continuous light alternate with dark winters, producing extreme seasonal temperature variations. This causes mean annual temperatures to rise in the polar regions, frost and ice to sublime, the atmospheric pressure to rise, greenhouse heating to increase, dust storm activity to increase, more dust to enter the atmosphere, and the climate to grow milder. In contrast, at times of low obliquity, mean annual polar temperatures and seasonal temperature fluctuations are at a minimum, causing carbon dioxide and water to freeze out of the atmosphere. As a result, the atmospheric pressure plummets, greenhouse heating decreases, dust storm activity subsides, the atmosphere becomes clearer, and the climate turns colder and drier. During periods of low obliquity, which pole will be the coldest is controlled mainly by topography instead of the dust cycle. During these periods, perennial carbon dioxide ice caps form because of low polar temperatures. Considering the current topography, during the minimum obliquity the South Pole should be about 6°C (11°F) colder than the North Pole (-124°C [-191°F] in the north versus -130°C [-202°F] in the south), making it the dominant perennial ice cap.

A quantitative connection between the timing of changes in obliquity (and the other motions of Mars) and the development of layered deposits has yet to be established and will be extremely difficult to prove for three reasons. First, the ages of the layers are not known with the precision needed for comparisons with the timing of the astronomical cycles. Second, obliquity changes in the orbit of Mars may occur in cycles that vary widely in duration and timing. Careful studies of the gravitational interactions of Mars with other planets suggest that the evolution of the obliquity of the Martian orbit may be chaotic due to a gravitational resonance with Venus. If this is the case, the history of the Martian obliquity can only be extrapolated several million years back. Third, materials such as the layered deposits that accumulate around the poles can change the mass balance of Mars, causing it to wobble. This influences the rate of precession, inducing polar wandering and creating a whole new climate regime. This effect is certainly real. Currently, there is a slight wobble in the spin of Mars caused by the asymmetric distribution of layered terrain and ice around the poles.

Are climate changes unique to Mars? The answer to this is a confident no. Ask any wooly mammoth. Earth's ice ages are a prime example of periodic swings in climatic conditions on planets. For Earth these climate changes have been traced to changes in its obliquity. Though periodic changes in Earth's obliquity have been known for many years, only recently has a quantitative connection been made between these changes and the timing of ice ages. Changes in Earth's obliquity are much less pronounced than those of Mars because Earth's moon helps to stabilize motion. Distinctive types of sea floor sediments produced as a result of ice age environments have been dredged up and dated. These sediments were deposited in cycles that exactly follow the timing of periodic changes in Earth's obliquity. As a result, it is tempting to speculate that periodic changes in the motion of Mars also cause periodic changes to its climate and that these changes have produced polar layered deposits and other terrain.

Comet Trickery

Comets may have led us down a road of false understanding about the evolution of the Martian atmosphere. Much of our understanding of the evolution of the atmosphere and climate comes from the study of chemical anomalies that are expressed as differences between the relative abundance of gases found in the current atmosphere and the relative abundance of gases initially incorporated in Mars from the solar nebula. Tobias Owen of the University of Hawaii and his colleague have warned that comet impacts might potentially add new volatile materials to the Martian atmosphere. Whether such materials have been added to the Martian atmosphere is particularly critical to our ability to work out its history. Because the composition of volatiles in comets may be very different than the composition of those incorporated into Mars as it accreted, addition of substantial amounts of cometary materials would contaminate the Martian atmosphere with gases of different and unknown composition. As a consequence, if these added volatiles are not taken into account, hypotheses of the evolution of the Martian atmosphere based on an assumed starting composition would be seriously flawed.

Generally scientists agree that early in solar system history a huge amount of volatiles was delivered to Mars by collisions with low-velocity comets and that these volatiles were quickly lost, blown off by the formation of huge impact basins. Most scientists also agree that comets have continued to pepper the planets with volatile-rich debris throughout the rest of solar system history. However, some scientists feel that comets probably have contributed little or nothing to the current Mars atmosphere because they are high-velocity objects. As a result of the high velocities, their impact energy is so great that the volatile materials they carry are automatically blasted into space during their collision with Mars. Instead of the atmosphere being derived from comets, these scientists generally assume that the current Martian atmosphere was formed mainly by escape of gases from the interior (during episodes of volcanism or tectonism). Consequently, the current atmosphere reflects the composition of the nebula from which Mars accreted.

Not all scientists agree with this assumption and, like Tobias Owen, think that Mars may have retained some of the volatile materials carried by the comets that have slammed into the planet over the past 3.8 billion years. As a result, these scientists hold that these comets brought with them a significant contribution of alien gases to the current Martian atmosphere. As evidence that this is possible, these scientists point out that impact events typically produced ejecta with a wide range of velocities, including that of the impacting projectile. Consequently, a portion of the volatile-rich ejecta from comets should be moving at well below the escape velocity of Mars and should be captured. They also offer as evidence the recent detection of ice in the soils of the permanently shadowed regions at the poles of the Moon. They suggest that the only reasonable source for this ice is captured gases delivered by comet impacts. Consequently, these scientists reason that if comets can deliver water to the Moon and it stayed, then comets can deliver volatiles to Mars that will also stay to be incorporated into the atmosphere.

The SNC meteorites also may hold clues to this question. A careful analysis of these rocks has shown that the relative abundance of oxygen isotopes found in their minerals is different from the relative abundance of oxygen isotopes found in Martian water trapped in the same SNC meteorites. Part of this difference is expected, caused by atmospheric escape processes. However, this process cannot account for the magnitude of the difference. Two possible explanations are likely for this difference: there is an additional loss process in operation in the atmosphere that we have not taken into account, or there were two distinctly different sources for the oxygen found in these meteorites. If the latter is the case, then one of these sources, reflected in the minerals, could be from primordial Mars, and the other source, reflected in the water, could be from comets. Until documented samples from both Mars and the comets are brought back for detailed analysis, this question likely will be unresolved.

Chapter 7

Searching for Martians

Searching for Martian life is a complex and puzzling problem. In the 1960s, when NASA first began to consider searching for life on Mars, no one really knew what to expect Martian life to be like. What do you look for if you don't know what you are looking for? Moreover, how do you look for it?

Since these questions were first asked, NASA has made a first attempt to find life on Mars. During the same time, science has undergone a revolution in the understanding of how life originated on Earth, how it has evolved, its great diversity, and the wide range of environments to which it has managed to adapt. This new knowledge has altered our view of what to look for and how to look for it. In this chapter, we will explore where the search for life has taken us, what we have learned, and where it may take us in the future. We will begin by asking why we search for life on Mars.

Why Search for Life on Mars?

To a large degree, Mars has been the prime target of exploration because of its potential for harboring life. What makes us so interested in searching for life on Mars? Finding it would have little or no economic value and probably would make little difference in most of our lives. No matter, society continues to be fascinated with finding life, especially on Mars.

Bruce Jakosky, an astrobiologist at the University of Colorado, has offered that one of the main reasons we search for life ultimately stems from our need to understand our own origin. He argued that "the drive to understand the distribution of life in the universe ultimately connects to our desire to understand the nature of the world around us and of the interactions between us and that world, and thereby to understand more about ourselves. Exploration of space, therefore, becomes one more means, along with exploration of the humanities, literature, and the arts, of understanding what it means to be human."

Jakosky continued, "In this context, finding microbes would be just as significant as finding intelligent life for what it tells us about the nature of life and its distribution in the universe. Ultimately, it points us toward 'looking for ourselves,' in terms both of understanding the origins of our own society and of finding out if there are comparable beings or societies elsewhere. As with the discoveries by Darwin and Copernicus, one cannot easily point to how such a discovery might affect one's day-to-day activities, even while the end result would be profound."

Added to our need to know our origin, the many myths, legends, and fictional accounts of exotic life-forms on Mars have fueled a fascination with Mars. The once-popular notion that Mars is an abode of intelligent life was sensationalized by its most ardent supporter—Percival Lowell. Lowell argued forcefully and eloquently that Mars was home to heroic beings struggling to survive in a dying arid world. Lowell was so convincing that his influence lived on, even though he was incorrect. In spite of being wrong, Lowell inspired such science fiction writers as Edgar Rice Burroughs, Ray Bradbury, H. G. Wells, and Arthur C. Clark to spin many of their stories with Mars as the backdrop. It was these and other writers of science fiction who perpetuated and popularized Lowell's views of Mars as home to an abundance of different types of creatures. Their adventure stories thrilled the imagination of several generations of readers. Is it any wonder that their readers long to know the inhabitants of Mars?

It should also be remembered that we search for life on Mars because, besides Earth, it has the greatest likelihood of any place in the solar system to harbor life. Mars has the most Earth-like surface environment of any other planet in our solar system, suggesting to many scientists that if life originated in Earth's environment, then life should also be expected to have developed on Mars.

Driven by these motivations, in the 1960s NASA began to plan the bold move of landing vehicles on Mars with the explicit goal of finding life. The result was the Viking Program, the most complex and expensive planetary missions ever flown. In the next section we will see how Viking approached the search for life on Mars, what it found, and what it did not find.

Viking

In 1976 the landing of Viking 1 and 2 marked the first, and so far the only, time spacecraft have left Earth with the expressed goal of finding life. Viking carried the most controversial payload ever prepared by the space industry. At a cost of $55 million each (in 1975 dollars), the two life detection experiments were the most expensive and important investigations on board the Viking landers.

What made these experiments so expensive? They required complex instruments never before used coupled with the very small payload capabilities of the landers. As a result, the science payload for each Viking lander had only 0.08 m³ (0.03 cubic feet) of space and 15 kg (33 pounds) of mass allocated for these two experiments. The Viking science teams that had the job of redesigning enormous laboratory instruments to fit on these tiny landers often joked that their job was "stuffing room-sized laboratories into a shoe box." Amazingly they succeeded.

From the beginning, it was clear that Viking had the difficult task of looking for something that could not be confidently defined. The Viking biologists worried about how they would approach this. After considerable debate, they concluded that the only sound approach was to assume that life on Mars would be similar in fundamental ways to life on Earth. In particular, they assumed that Martian organisms would be composed of organic molecules and that these organisms would employ the same metabolic processes used by the microbes common on Earth's surface. Reasonably so, because organic molecules provide the most diverse and flexible building blocks of life. Added to this, nearly all organisms known at the time Viking was designed used the same chemical pathways to power their metabolic processes. Considering how little we know about life, at the time this seemed a reasonable approach, though in hindsight it was a bit naive.

The detection of life by Viking was ultimately dependent upon two experiments: the biology (Figure 100) and molecular analysis experiments (Figure 101). Results from the biology experiments would detect the presence of organic life through its metabolic

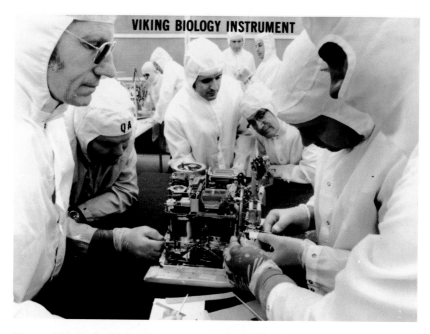

Figure 100. Technicians examining the Viking biology instruments. The instruments were used in the pyrolitic release, gas exchange, and labeled release experiments. Did any of these instruments detect life? (Courtesy NASA)

processes, and the molecular analysis experiments would determine if any organic compounds had existed in the past. In an attempt to detect the greatest range of possible organisms, the Viking biology experiments tested for evidence of metabolic processes under different sets of environmental conditions. Gerald Soffen, Viking project scientist, summarized this approach: "It was decided to send a set of biological tests that range in their environmental setting from a totally aqueous milieu, rich in organics, to a Mars-like environment with no water or any other additive."

The biology experiment consisted of three different experiments: the pyrolitic release experiment, the gas exchange experiment, and the labeled release experiment. Each was designed to detect aspects of metabolic processes associated with common living organisms. The pyrolitic release experiment exposed dry Martian soil samples (sterilized by heating carbon and carbon dioxide) to radioactive carbon; it was expected that any carbon-based organisms existing in the soil

Figure 101. The Viking molecular analysis instruments tested for organic molecules. These instruments included a gas chromatograph, a mass spectrometer, a sample oven and associated electronics, effluent divider, and gas separator. These failed to measure organic materials on the surface of Mars. (Courtesy NASA)

would consume the radioactive carbon as part of their metabolic process. Consumption of this radioactive carbon could be measured and would indicate if any were used by an organism. In the gas exchange experiment, nutrient-rich broth nicknamed "chicken soup" was added to prehumidified Martian soil with hopes that any microbe present would dine on the soup and give off gas by-products. The chemical composition of the gas by-products was continuously monitored for signs of biological activity. The labeled release experiment added several nutrients, in which radioactive carbon atoms were substituted for original carbon atoms, to the soil. If the nutrients were eaten by organisms, then carbon dioxide would be released by the organisms and the radioactive atoms could be detected.

The molecular analysis experiment took a different approach. It was designed to detect organic matter in the surface soil. Soil samples were baked in a tiny oven at high temperatures to drive released gases into the chromatograph, where the molecules were separated into their component parts. The parts were then identified as they passed through a mass spectrometer. The mass spectrometer was capable of detecting organic molecules at the parts-per-million level and even smaller concentrations of some molecules at the parts-per-billion level.

Though not designed as life detection experiments, the other instruments on Viking could detect life—under the right circumstances. Gerald Soffen joked, "Many of the Viking instruments could have detected life. The orbiter camera could have seen cities or the lights of civilizations. The infrared mapper could have found an unusual heat source from concentration of life forms. The water vapor sensor could have detected watering holes or moisture from some great metabolic source. The entry mass spectrometer could have identified gases that were wildly outside the limits of chemical equilibrium (as oxygen is on Earth). Seismometers could have detected a nearby elephant."

From the beginning, Viking biologists were worried about both the protection of the Martian environment from any hitchhiking terrestrial organisms carried aboard the landers and the possibility that attempts to detect Martian life could be thwarted by the detection of these terrestrial organisms. With such sophisticated and sensitive instruments aboard, it was clear that Viking was capable of detecting such organisms.

Protecting Mars

As with any life detection mission planned to go to Mars, NASA regarded it essential that any organism detected by Viking be Martian and not an Earthling stowaway. It was also deemed essential that Mars not be contaminated with Earth organisms that could grow on Mars to either be mistaken later for Martian life and/or act as a plague to displace any indigenous life-forms.

By some accounts, one-fourth of the project costs were expended to ensure that Mars would not be contaminated by organisms carried on the spacecraft. Both the orbiters and landers were cleaned extensively. Because the landers were to go to the surface of Mars, they also underwent additional decontamination measures. They were disassembled, placed in a special oven at microbe-killing temperatures for 48 hours, and heat sterilized. Then they were reassembled and sealed in a sterile capsule. This sterilization procedure required that special components be developed for Viking that could withstand such temperatures.

However, in spite of international agreements against contamination of Mars and the best efforts of all Mars missions to sterilize their spacecraft, including Viking, each has begun its voyage carrying living microbes. Complete sterilization by any known means (e.g., heat or highly toxic gas) is impossible without also destroying the spacecraft. Consequently, sterilization measures can only reduce the number of organisms clinging to the spacecraft. Although they were not entirely eliminated, these measures reduce the probability that hitchhiking microbes could survive and reproduce in the harsh Martian environment. But experiments have shown that there is no doubt that some of these microbes surely were alive when each of the probes landed.

The contamination of Mars is possible or even probable, although survival of hitchhiking organisms after landing may be a very different story. The solar radiation and chemical environment would quickly destroy them if they tried to venture from protected areas of the spacecraft. Both of these environmental factors break down organic molecules, the stuff of which terrestrial organisms are built.

Based on the results from Viking life detection experiments discussed in the next section, Viking clearly did its job well enough to prevent any biologic contamination it carried from yielding false positive results.

Puzzling Results from Viking

Shortly after Viking 1 touched down, the sampler arm scooped up some Martian soil and delivered it to the life detection instruments. One by one, these tiny sophisticated instruments began to churn out their results. And were these results remarkable! Two of the biology experiments produced positive responses. However, only one, the labeled release experiment, produced a response that satisfied the original Viking mission criteria for a positive result. It showed an initial burst of radioactive carbon dioxide followed by a slow, continuous release of gas suggestive of biological activity. The pyrolitic release experiment showed a less conclusive response: only a minute rise in radioactive carbon was detected. This rise occurred even after heat sterilization at 90°C (194°F) for 2 hours, suggesting that this rise was not due to a living organism.

In a spectacular chemical reaction, when the first samples placed in the gas exchange experiment were prehumidified, they produced a rush of oxygen, although later, when the "chicken soup" was added to the same sample, only a small amount of carbon dioxide was released, with no sign of an additional release of oxygen. The experiment was repeated three times with the same results. Most scientists began to suspect that the results from this experiment and those from the pyrolitic release experiment could most easily be explained as a chemical reaction in the Martian soil.

At odds with the possibility that life had been detected, the molecular analysis experiment detected no organic molecules in the soils. This finding was stunning, considering that small background levels of organic molecules were expected on Mars. As on Earth, a slow but continuous rain of meteorites falls on Mars. Many of these meteorites (e.g., carbonaceous chondrites) contain nonbiogenic organic molecules that should have been detected. The rain of these meteorites scatters organic molecules around the surface of Mars and should make up a detectable component of the soils. In addition, because of the nature of the sterilization process used to decontaminate the spacecraft the lander unavoidably carried another source of organic molecules in the form of small levels of biologic contamination. These also were not detected. For Viking scientists this result was the most puzzling. Reflecting on these results, Gerald Soffen declared, "The biggest single intellectual error that was made in the entire Viking program was that all of us absolutely knew there would be organics on Mars—we just knew it. None of us dreamed there would be no organics, or that we would not find any."

As these results came in, the Viking science teams worked frantically to explain the observation in context, knowing that the world was waiting for word on the most important question—is there life on Mars? "This was such an unusual time on the project because we were always in the analysis cycle," said Norman Horowitz, designer of the pyrolytic release experiment. "Everyone was constantly looking over our shoulders at scientists who were trying to work in a normal way in an abnormal environment. Nobody wanted to be wrong in public on a question as important as that of life on Mars." In spite of the pressure of the situation, the teams methodically went about the task of analyses and interpretation of the results. They asked themselves: what did such puzzling and seemingly contradictory results mean? Different team members had very different answers to this question.

Did Viking Find Life?

As the results were analyzed, it quickly became clear to most of the members of the Viking science teams that the prospect of finding life on Mars was dimming. Harold Klein, a Viking biology team member from NASA Ames Research Center, summarized what most Viking scientists had concluded: "for each experiment, except for the labeled release experiment, we conclude that there were no organisms present within the limits of detection of the experiments and that the observed reactions from these were the result of nonbiological phenomena."

Norman Horowitz shared this view and felt that the hope for finding life on Mars vanished with Viking. In considering the ultimate value of the Viking results, he suggested that the Viking results provide a new perspective on our own home, Earth. He commented, "It now seems certain that the Earth is the only inhabited planet in the solar system. We have come to the end of the dream. We are alone—we and the other species that share the planet with us. If the Viking findings can make us feel the uniqueness of the Earth and thereby increase our determination to prevent its destruction, they will have contributed more than just science." He added that "For some, Mars will always be inhabited, regardless of the images. You do not have to search far to hear the opinion that somewhere on Mars there is a Garden of Eden—a warm, wet place where Martian life is flourishing. This is a daydream. The Garden of Eden would reveal itself in photographs by a permanent water cloud above it and by snow on the ground. Nothing like this has been seen and it is most unlikely that such a place exists or can exist on Mars."

What about the results from the labeled release experiment? After all, they met the criteria for detection of life. Harold Klein explained, "On the basis of all of the experiments performed to date, the labeled release experiment, unlike the other biological experiments, yielded data which met the criteria originally developed for a positive. On this basis alone the conclusion would be drawn that metabolizing organisms were indeed present in all samples tested. Can we believe such a conclusion? Clearly, we must be wary of this in the face of information indicating that all of the samples tested yielded oxygen in the gas exchange experiment upon introduction of water. The evidence for strongly oxidizing chemicals in these samples is quite convincing." To most Viking scientists this meant that the labeled release results could most easily be explained by chemical reactions in the soil and not life.

In spite of the rationale offered by the other Viking scientists that life had not been detected, Gilbert Levin, the designer of the labeled release experiment, remained optimistic. He was confident that life exists on Mars and that his experiment had detected it. He responded to suggestions that his experiment had not found life by declaring that its positive results had, indeed, satisfied "the pre-flight criteria for the detection of life approved by NASA, the National Academy of Sciences and the Space Science Board." He further countered his detractors by arguing that the life detection criteria "were fully demonstrated" and complained that "Yet the results were deemed unconvincing due to the lack of supporting evidence from other experiments onboard the Landers. We were informed by NASA that it had selected the three experiments because they each tested different models of Mars, and were there truly life on Mars, only one experiment should return a positive response."

On this point of a positive response by one experiment proving life, Gerald Soffen offered a reminder of another ground rule of Viking concerning life detection. He stated that, "The various biology experiments under consideration were all tests for some biochemical response of a Martian soil sample inoculated with some kind of medium. No single test was considered an adequate first test for life."

To most Viking scientists, the failure to detect any organic compounds in the soil meant no life. Before the landing on Mars, these scientists had assumed that organic molecules would be present on Mars. That these molecules were not found was, according to Soffen, the single most important discovery made by Viking. Faced with the results of the molecular analysis experiment, most Viking science team members concluded that something must be destroying these molecules. "In some ways you could say that Mars is self-sterilizing," Soffen explained. He speculated that Mars "has an oxidizing surface and when a meteorite comes in and organics fly off and land on the surface, the iron peroxide on the surface chews up this natural organic material and wipes the surface clean. It's really incredible in a sense. Mars is a lot like Earth; it has an atmosphere, mountains, valleys, and clouds. But it has an oxidizing surface that keeps it clean and sterile." Adding to the effects of oxidants in the soils of Mars, ultraviolet rays from the Sun easily break apart organic molecules, killing all living things dwelling on the surface.

Gilbert Levin and his team were quick to acknowledge that the lack of organic molecules created an important inconsistency between the interpretation of their results and the results from the other experiments. Levin declared that "The molecular analysis experiment found no organics in the Martian soil and this created a big problem. The result—there were many attempts to explain the labeled release experiment results without invoking the possibility of life on Mars." He and his team felt strongly that such an explanation was not necessary in light of the results from their experiment and that the Viking criteria for life detection had been met by their results.

Some researchers have gone to great lengths to explain how life can inhabit the surface of Mars without leaving a trace of organic molecules. Some have theorized that the Viking instruments actually found life but the population density of the Martian organisms was below the detection level of the molecular analysis experiment. Alternatively, it has also been suggested that the Martian organisms are composed of a type of organic molecules not detected by the experiment or that the molecular analysis experiment failed to work properly on the planet's surface. According to Norman Horowitz, this is a form of the "blue unicorn" theory. This theory holds that a blue unicorn is living in a cave on the Moon. The theory cannot be disproved because its author provides the unicorn with whatever attributes are needed to survive on the Moon. On Mars these would include the ability to live without water or any other solvent and immunity from the processes that destroy all other forms of organic matter on the surface.

It may be a long time before this controversy is settled. Most of the scientific community considers the likelihood that life inhabits the surface of Mars to be very small. In considering the Viking results, Gerald Soffen summarized his feeling about what he and most of the science community thought about the search for life on Mars after Viking:

I have devoted 20 years to thinking about this question. In my role as Viking project scientist during the primary mission, I began with a very optimistic view of the chances for life on Mars. I now believe that it is very unlikely, but one doubt lingers: we have not visited the polar regions. I have always believed that in the search for life, you must go where the water is. The permanent polar caps of Mars are frozen water and would act as a splendid "cold finger" of the planet to trap organic molecules. The oxidizing agents should be absent. . . . And who knows what those lucky future explorers of Mars will find there.

Thus, in spite of carrying the most sophisticated science payload yet flown by NASA, the Viking landers found little to suggest life. However, because Viking could only perform a few simple experiments, its results have not closed the question of life on Mars. The Viking results have merely lowered the probability that life exists there.

Our Changing View of the Search for Life

When Viking was conceived and developed, biologists were unaware of the organisms that thrive in extreme environments on Earth. These organisms violate Viking's assumption that all creatures use the same metabolic processes. Since Viking, primitive organisms, called extremophiles, have been discovered thriving in places previously thought uninhabitable, such as boiling water in hot springs, around volcanic vents beneath the sea, in deep oil reservoirs, and in rock hundreds of meters below the surface. Some of these organisms would flunk the Viking biology tests. They do not use the same sources of energy to support their metabolic processes that Viking assumed was common to all organisms. Undoubtedly, if the Viking biologists had been aware of the surprising metabolic diversity of terrestrial organisms, the Viking approach to detecting life surely would have been different.

Genetic studies indicate that the organisms that live in terrestrial extreme environments are closely related to the first life-forms on Earth and may provide the best starting point for basing a search for life on Mars. Many biologists now think that life on Earth originated with extremophiles in hydrothermal vents (hot springs generally heated by igneous activity) where warm, mineral-rich waters provided a chemical and physical environment rich in the ingredients necessary for life. This hypothesis has dramatic implications for the possibility of life on Mars. Like Earth, Mars has also had a long volcanic history and abundant subsurface water needed to produce this same environment. Biologists question that if life originated on Earth in this type of environment, then why should it not have originated on Mars?

Kenneth Nealson of the Jet Propulsion Laboratory has done extensive work on extremophile organisms and based on this work considers the energy flow associated with life processes to be one of the prime measures of life. He and his colleagues have found that there are extremophile organisms commonly filling niches left behind by the oxygen-using, carbon-eating organisms common on the surface of Earth. These organisms are not oxygen using or carbon eating. He has pointed out that "Such extremophily forms a backdrop by which we can view the energy flow of life on this planet, think about what the evolutionary past of the planet might have been, and plan ways to look for life elsewhere, using the knowledge of energy flow on Earth." He explained that the primitive extremophiles "are metabolically very diverse; they are able to use almost any energetically useful chemical energy that is abundant on Earth. Evolution and competition have undoubtedly driven these ingenious chemists (the extremophiles) to develop methods for harvesting virtually every worthwhile corner of the chemical market, including both organic and inorganic energy sources of nearly all kinds."

With these new insights into the nature of life, the paradigm that originally guided the search for life with Viking has shifted considerably, not only because of what was found on Mars, but also what has been found on Earth. Bruce Jakosky reflected on the new paradigm:

Based on observations of the Earth's geologic record, life on Earth originated rapidly after it became possible for life to exist. This indicates that an origin is likely a direct and straightforward consequence of chemical reactions that can occur in a natural planetary environment. More than any other observation, this suggests that life might originate on any planet that had a similar environment, and that life could be widespread throughout the universe. Indications today are that the environmental prerequisites for life include only the presence of liquid water, access to the biogenic elements, and a source of energy that can drive chemical disequilibrium. These are not terribly stringent requirements. In our solar system, as many as a half-dozen planets or satellites may have met these conditions at one time and could, therefore, have sustained life.

Victor Baker of the University of Arizona pointed out that these findings have direct relevance to life on Mars because "The extensive hydrosphere implied by past aqueous activity on Mars may only be extant ground ice in the thick permafrost zone and as underlying groundwater. Yet, this is the type of environment in which the extremophiles progenitors of Earth's biosphere probably evolved. Indeed, early Mars provided an arguably better habitat for the inception and incubation of early life than did Earth. Episodic, brief episodes of aqueous activity on the Martian surface may have exposed this biosphere to produce possible fossil indicators of its existence." He speculated that, remarkably, under the right circumstances the biosphere could have affected the climate as much or more than the climate affected it. He suggested, "a deep subsurface containing methanogenic archaea (primitive organisms that use and produce methane as part of their metabolic processes) could have produced methane that accumulated beneath a growing Martian cryosphere. The result could destabilize the Martian cryosphere and perhaps change the climate on short time scales."

To biologists the discoveries of extremophiles and what they mean to the possibility of life on Mars are exciting. They have provided a deeper understanding of how living organisms work and what constitutes life. This new knowledge calls for new approaches in our continued search for those elusive Martians.

A Gift from Outer Space

Unexpectedly, in 1996 the search for life on Mars took a huge step forward. As part of their research on meteorites at NASA, David McKay of NASA's Johnson Space Center and his colleagues set about examining the globules of carbonate minerals (Figure 102) found in cracks in meteorite ALH84001 (Figure 103). This meteorite is 4.1 billion years old and, like the SNC meteorites (named after the three major types of igneous meteorites: shergottite, nakhlite, and chassigny), shows most of the same evidence to indicate that it came from Mars. McKay and his team thought these minerals were deposited by warm mineral-bearing water as it circulated through the fractures (an environment similar to that found in hot springs on Earth). What these scientist discovered inside the carbonate material led them to a startling conclusion—the rock contained evidence

Figure 102. Yellowish carbonate globules are found in cracks in Martian meteorite ALH84001. These globules are thought to have been deposited from mineral-rich warm water that circulated through the Martian crust. Some scientists think that these globules contain evidence of organisms that lived in the water. The view is about 0.5 mm (0.002 inch) wide. (Courtesy NASA) (PIA00290)

of possible biogenic activity on Mars. According to this team, they based their interpretation on four main lines of evidence: "the presence of carbonate globules which formed at temperatures favorable for life, . . . the presence of biominerals (magnetites and sulfides) with characteristics nearly identical to those formed by certain bacteria, . . . the presence of indigenous reduced carbon within Martian materials, and . . . the presence in the carbonate globules of features similar in morphology to biological structures." McKay's team acknowledged that "Each of these phenomena could be interpreted as having abiogenic origins." The team suggested that "the unique spatial relationships indicated that collectively, they recorded evidence of past biogenic activity within the meteorite."

Based on this evidence and their confidence in their interpretation, in 1996 the McKay team announced that they had discovered evidence for ancient life in the rock. This announcement set off one of the most vigorous, and at times acrimonious, debates in recent scientific history. This was such an important discovery that the science community wanted to be absolutely certain that it could stand up under the most rigorous scrutiny. The science community echoed what Carl Sagan had said about any claim that was considered outrageous, "Extraordinary claims require extraordinary proof." As a

Figure 103. Two views of Martian meteorite ALH84001, which may contain evidence of primitive life on early Mars. This rock formed on Mars over 4 billion years ago, was blasted off by a giant impact about 16 million years ago, and fell to Earth in Antarctica 13,000 years ago. The small cube with the M is 1 cm (0.4 inch) across and is for scale. (Courtesy NASA) (PIA00289)

result, ALH84001 has become the most studied rock in history, as the scientific community has torn it apart searching for this "extraordinary proof."

The controversy over evidence of life in this meteorite still rages on and may well continue for some time. McKay and his colleagues, reflecting on their findings, suggested that "if you believe the data commonly provided as evidence for the oldest life on Earth, then because the data is so similar for Mars, you should also believe that ancient primitive life must have also existed on Mars." Opposing this view, many scientists do not accept the premise that "similar" is identical to "same." Nor do they accept that a chain of evidence added up to prove life in this rock, when each of the links could be interpreted to be of abiogenic origin. As these scientists have dug into ALH84001, the foundations of this evidence have started to crumble. Most of the evidence has been shown to be either a result of a nonlife process or equivocal with regard to being solid evidence for life. For example, the tiny wormlike features that appear in the electron microscope picture of ALH84001 (Figure 104) may have been a product of the sample preparation process and are not microfossils. Even more remarkable, it was found that microfossils of

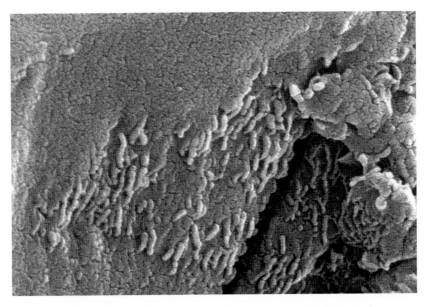

Figure 104. Bacteria-like features found in ALH84001. These tiny features, once thought to be microfossils of bacteria, may have been produced as a result of the sample preparation process. (Courtesy NASA) (PIA00284)

Earth organisms can form inside rocks in a relatively short time. Researchers found microfossils of terrestrial bacteria in the Nakhla meteorite. This meteorite was collected when it fell to Earth nearly 100 years ago and then immediately deposited in a museum collection. These microfossils must have been formed in less than 100 years, thought to be created by reactions between the rock and humid air that produced mineral coatings deposited on dead terrestrial bacteria. However, one line of evidence still stands: the type of magnetite (Figure 105) found in the carbonate globules has been produced only by living organisms (either in nature or in the laboratory).

Though the discovery of McKay and his team may have taken numerous body blows and is on the ropes, it is not out yet. The magnetite must be explained. But even if this discovery is eventually shown to be a false alarm, it has sharpened the awareness of NASA and the science community about the difficulties they will face in proving anything found in samples brought back from Mars as signs of life.

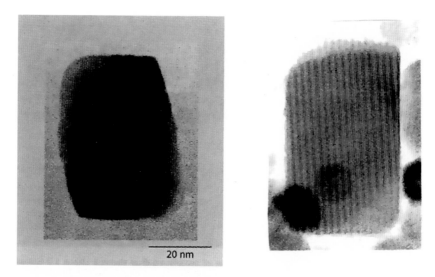

Figure 105. Magnetite crystals from Mars and Earth. Transmission electron microscope images of magnetite grains: *(left)* found within the rims of carbonate globules from ALH84001 and *(right)* produced by an Earth bacterium, shown at the same scale. Stripes in the Earth crystal are layers of iron atoms. Both of these grains meet all the criteria for magnetite produced by bacteria. Will other ways be found to produce this type of magnetite grains? (Modified from K. L. Thomas-Keprta, D. A. Bazylinski, J. L. Kirschvink, S. J. Climett, D. S. McKay, S. J. Wentworth, H. Vali, E. K. Gibson, and C. S. Romanek, Elongate prismatic magnetite crystals in ALH84001 carbonate globules: Potential Martian magnetofissils, *Geochimica et Cosmochimica Acta,* 64:4049–4081, 2000)

Cosmic Hitchhikers: Friends or Foes

For a moment, imagine: what if Mars had abundant life and these life-forms could survive the trip aboard the SNC meteorites from Mars to Earth? Could they survive once they were here? How would these life-forms that grew out of the Martian environment interact with terrestrial organisms? Would they be benign or pathogenic?

One view holds that there may be little danger that Martian organisms would be pathogenic. This view assumes that Martian organisms carry a different genetic code than terrestrial organisms, a result of development in different environments. Different genetic codes would produce very different biochemical pathways and provide no way for organisms to infect one another.

Other views argue that Martian organisms might be genetically and biochemically very similar to terrestrial organisms. Some scientists argue that clay minerals may have acted as blueprints for building complex organic molecules that themselves are the building blocks of life. If this is the case, then organisms from both planets could be biochemically and genetically very similar. This similarity might allow terrestrial and Martian organisms to infect or prey on each other.

To other scientists it is not clear whether a difference in genetic code and biochemistry has a bearing on this question. They argue that there appears to be no known reason for an infective agent or predator to require the same biochemical pathways or the same macromolecules as its victim. They suggest that organisms need not be similar to infect or prey on each other.

For Mars, this debate may be moot. We may already know the answer. We know for certain that Martian SNC and lunar meteorites have fallen on Earth. If microbes exist on Mars, then contamination of Earth by unaltered materials blasted off the surfaces of planets may have occurred many times over the history of the solar system. Have they carried hitchhiking organisms inside them? What has been the effect? All known organisms on Earth came from one original parent organism, suggesting to some scientists that no organisms from the outside have been introduced. Other scientists counter this and suggest that it does not preclude that the first organism was planted here from elsewhere (i.e., Mars).

The reverse of this situation must also be true, although it is more difficult to transfer materials from Earth to Mars by the same process. There is no reason that unaltered Earth meteorites carrying Earth organisms could not have also fallen on other planets, such as Mars. It should be kept in mind that there are types of common Earth microbes that could easily survive the ejection process and harsh conditions encountered in space during the long journey to Mars. Has Earth contaminated Mars? Should we expect to find the descendants of these cosmic hitchhikers hiding in protected areas on Mars waiting to be discovered?

Chapter 8

The Moons of Mars

The two tiny moons of Mars, Phobos and Deimos, were discovered in 1877 by Asaph Hall at the U.S. Naval Observatory in Washington, D.C. But their presence had been predicted two and a half centuries earlier by Johannes Kepler. The prediction was erroneously based on an apparently simple sequence of increasing numbers of moons of successive planets. The sequence added one moon per planet, starting with Venus with no moon, the Earth with one, Mars with two, a suspected missing planet between Mars and Jupiter with three, and Jupiter with four. At the time of Kepler's prediction, Galileo had just discovered the four large moons of Jupiter. But we now know that Jupiter has many more small moons.

Kepler's prediction was known to Jonathan Swift, who mentioned it in *Gulliver's Travels.* He described the inhabitants of the flying island of Laputa as having "discovered two lesser stars or satellites which revolve around Mars." This sparked the popular belief that Mars had two tiny moons and caused considerable interest in searching for them. For many years the moons of Mars escaped detection, probably because of their closeness to Mars, small size, and dark color. Several noted early astronomers (Herschel in 1783 and D'Arrest in 1864) had searched unsuccessfully for the satellites of Mars, but it was Hall who first observed both satellites in the summer of 1877.

As is traditional after his discovery, Hall had the honor of naming these two tiny satellites. He wrote of his choice: "Of the various names that have been proposed for these satellites, I have chosen those suggested by Mr. Madan of Eton, England, via: Deimos for the outer satellite; Phobos for the inner satellite. These are generally the names for the horses that draw the chariot of Mars, but in the lines referred to (in the Fifteenth Book of the *Iliad*) they are personified by Homer, and mean attendants, or sons of Mars. He (Mars) spoke, and summoned Fear (Phobos) and Flight (Deimos) to yoke his steeds."

The orbits of both satellites were found to be nearly circular and close to the planet: Phobos orbits at a distance of about 9,378 km (5,627 miles) from the center of Mars and Deimos at about 23,459 km (14,075 miles). The orbital periods are expectedly short: 7 hours, 37 minutes for Phobos, and 30 hours, 18 minutes for Deimos.

At first the short periods were confusing to Hall and he commented: "For several days, the inner moon (Phobos) was a puzzle. It would appear on different sides of the planet in the same night, and at first I thought there were two or three inner moons. . . . To decide this point, I watched the moons throughout the nights of August 20 and 21, and saw that there was in fact but one inner moon which made its revolution around the primary in less than one third the time of the primary's rotation, a case unique in the solar system."

Both satellites orbit in the same direction as the planet rotates. These motions, combined with the orbital periods of the satellites, mean that Phobos rises in the west and sets in the east twice a Mars day. Deimos rises in the east and sets two and a half days later in the west. Three days later Deimos rises again in the east.

The angle of inclination of the orbital planes of both the satellites to the Mars equatorial plane is small. For Phobos, it is 1.13° from the equator, and for Deimos it varies between 0.85 and 2.69° from the equator (the variation is caused by perturbations from the Sun's gravity). A combination of the close proximity of the satellites to Mars, the curvature of Mars, and the low angle of inclination of their orbits always puts both satellites below the horizon at high latitudes. Phobos is never seen from above 70° north or south, and Deimos is never seen from above 83°N.

The orbits of these satellites have been the source of considerable controversy. In 1945, after an analysis of the orbital motion of Phobos, astronomer B. Sharpless reported that its orbit was getting progressively smaller and soon Phobos would crash into Mars. Although Sharpless's conclusion was correct, it was found later to be based on inaccurate measurements of the orbit of Mars. Later more precise measurements confirmed that tidal effects were causing the orbit of Phobos to decay slowly. In about 100 million years Phobos will plunge into Mars, blasting an impact crater over 100 km (60 miles) across.

Measurement inaccuracies frequently resulted in bizarre ideas concerning Phobos. Some of these have the ring of science fiction. In 1959, using data found later to be inaccurate, I. Shklovskii of the Soviet Union in a study of the orbit of Phobos produced evidence that it must be hollow. He carried this finding to its logical end, concluding that for Phobos to behave as observed it must be a huge artificial satellite. This sparked popular interest in this satellite as being evidence for extraterrestrial intelligent life.

In 1969 Mariner 7 took the first pictures of the satellites from a spacecraft. Though of very low resolution, these pictures showed Phobos to be elliptical and only half as large as previously believed. Three years later Mariner 9 snapped numerous, relatively high-resolution images (about 200 m [656 feet] resolution) of both satellites. These pictures showed them to be heavily cratered, potato-shaped rocks. Since Mariner 9, the Viking orbiters, Mars Global Surveyor, and the Soviet Phobos spacecraft have built up our collection of pictures and spectra (though limited) of both satellites.

During the closest flybys of the moons, both the Viking and Mars Global Surveyor spacecraft flew so close to these satellites that extreme care was needed not to crash into them. These close flybys provided some of the most surprising and highest quality data about these satellites. In particular these flybys were so close that the moons' tiny gravity fields affected the path of the spacecraft enough to determine the mass of both moons. Knowing their masses and volumes (estimates from the pictures) allows the calculation of their densities. It was found that both moons are of remarkably low density, a little less than 2 g cm^{-3} (0.04 pounds/cubic inch). Many scientists were surprised at this value, expecting instead a higher density. Because these satellites are extremely dark-colored, these scientists expected that the moons were composed of the same dark materials as carbonaceous chondrite meteorites. These meteorites are a class of dark-colored, carbon-rich meteorites believed to originate outside the orbit of Mars. The density of these meteorites is about 3 g cm^{-3} (0.06 pounds/cubic inch).

Origin

The origin of Phobos and Deimos is unknown, but there are two currently popular theories. One theory holds that these bodies are captured asteroids. The other theory suggests that they are asteroid-sized pieces of debris left over from the initial accretion of Mars from the solar nebula.

Astronomers generally agree that small bodies (asteroids, comets, or other debris) that wander the solar system, if captured by a planet, would typically be trapped in a highly elliptical, inclined orbit. Consequently, when astronomers see a small body that has a circular and low-inclination orbit around a planet, they generally regard this as evidence that the small satellite was not a captured wanderer but a piece of debris left over from the formation of the planet.

Some scientists argue that wandering solar system bodies can be captured in an initially highly elliptical orbit and then later have their orbits circularized. This requires the delicate balance of many conditions to work, making many scientists uneasy about its validity.

The circularization of an initially elliptical orbit can be done through a mechanism akin to "aerobraking," used by modern-day space flight controllers to slowly circularize highly elliptical orbits of some spacecraft. Under the right conditions, if an asteroid was initially captured into just the right highly elliptical orbit around Mars, the lowest part of its path could have carried it repeatedly through the upper part of the early massive Martian atmosphere, slowing it by atmospheric drag and slowly circularizing its orbit.

Circularization of an orbit can also be done through a mechanism involving tidal forces tugging on the satellite. Depending on the orbital period and spin period of Mars at the time of their capture, it has been calculated that tidal forces could pull on these satellites in a way that would eventually drag them from their initial highly elliptical orbits into circular ones.

Though these mechanisms require special conditions to work, some astronomers suggest that there is evidence that these moons are asteroids captured when they wandered from the outer solar system: their density and composition. The densities of Phobos and Deimos are surprisingly low, too low for them to be made of the materials that formed Mars and even too low to be consistent with any class of meteorite. However, the density of the most volatile-rich type of carbonaceous chondrite meteorites is a much closer match and with the addition of water ice would be identical. Spectral measurements shows that these moons are composed of nearly black rock, similar to carbonaceous chondrites and asteroids found in the outer part of the main asteroid belt, though very different from asteroids found closer to the Sun or rocks found on Mars.

The counter argument to this holds that if Phobos and Deimos formed in place around Mars, then they should be composed of the same high-density material as the planet. To account for their low density, some scientists have speculated either that these moons are composed of a mixture of high-density Mars rock and low-density ice, trapped in their interiors, or that they contain enough pore space to bring their density down to the observed values. In addition, to account for the spectral properties of the bodies, these scientists argue that Phobos and Deimos are covered by a veneer of higher-density carbonaceous chondritic material delivered from the outer solar system, evidenced by the numerous impact craters.

Composition

The low densities, dark color, and spectral properties of these moons are the primary basis for estimates of their composition. Spectra made from Earth-based telescopes and by the Phobos mission indicate that the surface of these satellites is covered by dark, bland, waterless, rocky material. Scientists have searched the lists of known materials in the solar system (e.g., meteorite types) looking for what matches both the density and spectral properties of these satellites.

The best spectral match with these moons is a class of meteorite called ordinary chondrites. These are high-density, dark, dense, anhydrous (dry) iron- and magnesium-rich rocks. The high density of these materials appears to rule them out as being the same material that makes up Phobos and Deimos.

The spectra of carbonaceous chondrites are also nearly a match with the spectra of these moons. Unlike Phobos and Deimos, which show no evidence of water in their spectra, the carbonaceous chondrites are water-rich (up to 10–20 percent water). Taken at face value, this would rule out a carbonaceous chondritic composition for these satellites.

The spectra of these moons show many similarities with the spectra of asteroids in the outer asteroid belt. Though none matches exactly, all are dark, low-density, small bodies. To some scientists, these similarities suggest a common origin. Many argue that these moons originally may have formed in the outer asteroid belt and later wandered to their current location.

This absence of a match with any solar system material leaves scientists with a dilemma. There are several plausible solutions to this dilemma. For example, surface materials on these satellites may have lost their water, driven off by the shock and heat from the many small impact craters that produced their regoliths. Because spectral measurements see only components on the surface, only the anhydrous materials in the regolith would be detected and the water-rich rock inside would remain hidden. In this scenario, Phobos and Deimos could be made of low-density, water-rich carbonaceous chondrite.

Alternatively, these bodies could be composed of several different materials with characteristics that combine to yield the observed properties. Spectral mapping of Phobos by the Phobos spacecraft suggests that this moon is composed of several different materials. Scott Murchie at the Applied Physics Laboratory has analyzed these spectra and concluded that Phobos is composed of at least four (and perhaps more) different types of rock. He found that the spectra of the ejecta from the largest crater on Phobos, named Stickney (the maiden name of Asaph Hall's wife, who according to legend strongly encouraged him to head off to the telescope one more time on the night he finally discovered the two moons), are most like the spectra of ordinary chondrite meteorites. The rest of the surface of Phobos appears to be covered by regolith of a similar, though slightly different composition. He concluded that Phobos may be a mixture of low-density carbonaceous chondrite and high-density ordinary chondrite.

Some scientists think that one material likely to be found inside these satellites is water ice. A mixture of higher-density materials, such as chondritic rock, and a low-density component, such as water ice, is consistent with the low densities of Phobos and Deimos. If the materials in these satellites contained a considerable proportion of water ice buried deep inside (protected from evaporation), this would explain the apparent inconsistency between their low bulk density and the spectral data.

Other scientists argue that this is not necessary and that a solution to this dilemma may lie in the amount of open space and cracks generated by the continuous pounding of meteorites over the past 4 billion years. Impact crater formation events, especially large ones, would shatter these satellites, producing abundant pore space. The empty spaces substantially lower the overall density of the bodies. The lunar regolith provides a good example of the effects of pore space on density. Though lunar rocks have densities of about 3 g cm^{-3}

(0.06 pounds/cubic inch), the lunar regolith, derived from these rocks, has a density of only about 2 g cm^{-3} (0.04 pounds/cubic inch). This dramatic decrease in density is a direct result of the contribution of the vacant spaces between the individual rock fragments. In planets with high gravity fields (for example, the Moon), pore spaces are held shut by hydrostatic pressure. Unlike these planets, the gravity fields of tiny Phobos and Deimos are so weak that pore spaces in the deep interior can easily remain open. Considering the amount of such open spaces we know is generated in the crust of the Moon, this process must contribute to lowering the density of these satellites. How much it does is not known.

Considering the limited amount of data and what information it provides about the composition of Phobos and Deimos, most scientists think that these satellites are composed mostly of carbonaceous chondrite materials, with lesser amounts of ordinary chondrite mixed in.

Phobos

Phobos is a small, dark, heavily cratered rock, measuring only about 27 by 21.5 by 19 km (16.2 by 12.9 by 11.4 miles) (Figure 106). Its surface is uniformly covered with impact craters. The shape and size frequency of these craters are much like those of small craters on the Moon. Its largest crater, Stickney, is 10 km (6 miles) in diameter. It is currently impossible to estimate an age accurately for the surface of Phobos. However, its high density of craters suggests an ancient surface, similar to those on the planets and satellites found throughout the solar system that were formed during the epoch at the end of heavy bombardment about 4 billion years ago.

There is a complete range of states of preservation of craters from very old subdued pits to fresh craters with raised rims. The raised rims are probably due to structural uplift and not ejecta deposits. Most ejecta thrown from craters on Phobos escapes its small gravity field instead of producing secondary craters (the escape velocity for ejecta from Phobos is only 15 m/sec [49 feet/second] compared with 5 km/sec [3 miles/second] for Mars). Remarkably, even though the escape velocity of most of the impact ejecta blasted off this satellite exceeds 15 m/sec (49 feet/second) and it should be completely lost from the moon, it is not. Because Phobos orbits close to Mars, the gravity field of Mars is strong enough to capture most of the ejecta

Figure 106. This photomosaic was taken of Phobos from a distance of 612 km (366 miles). The large crater on the left is Stickney. Some of the grooves associated with Stickney are clearly visible. The crater at the bottom of the picture, Hall, is 5 km (3 miles) in diameter. (Courtesy NASA, Viking) (P20776)

from Phobos. Consequently, debris blasted from the surface of Phobos generally orbits Mars and ultimately crashes back into Phobos. This material collects on the surface, adding to the regolith. The same is true for Deimos.

Not all regolith accumulates through this capture mechanism. A few large blocks of ejecta, some house-sized, are observed sitting on the surface of Phobos on the rims of craters (Figure 107). It is generally thought that these are the parent craters for the blocks. If this

Figure 107. Blocks on the rim of the crater Stickney. The blocks are up to 10 m (33 feet) across and are thought to be materials ejected from Stickney during its formation. (Courtesy NASA/Jet Propulsion Laboratory/Malin Space Science Systems, Incorporated) (PIA01136)

assumption is valid, then these blocks must have been ejected at a velocity low enough for the weak gravity field of Phobos to capture them. There is also some evidence that some ejecta may have been deposited directly on the surface after being thrown from their parent craters. There appears to be a thin deposit of ejecta around the rim of Stickney.

Peter Thomas and Joseph Veverka of Cornell University calculated the amount of ejecta thrown from the observed craters on Phobos (assuming that most would eventually end up back on the surface of Phobos) and found that this debris should build up to make a regolith about 150 m (492 feet) thick. Their estimate is consistent with layering found about 100 m (328 feet) down the inside walls of several craters on Phobos. This layering is thought to mark the subsurface transition from solid bedrock to weak ground-up regolith materials. Likewise, hummocks and flat floors in other craters are thought to have been produced by this same transition. These features occur in craters with depths of between about 35 to 200 m (115 to 656 feet) on Phobos. The same types of interior features are observed in small lunar craters and are known to mark the regolith/bedrock transition.

The repeated formation of relatively large craters such as Stickney (10 km [6 miles] diameter), Hall (5 km [3 miles]), and Roche (5 km [3 miles]) are likely to have produced much of the regolith. These impacts would not only have excavated large volumes of rock but also would have left Phobos fractured throughout. The crater Stickney nearly took this process a disastrous step further. It has been calculated that if Stickney had been a few percentage points larger, it would have shattered Phobos, producing a number of smaller moonlets.

Though Stickney did not destroy Phobos, its formation still left its mark in the form of a moonwide system of grooves that radiates from this large crater. These grooves are most prominent near the rim of the crater (Figure 108). Some are as much as 700 m (2,296 feet) wide and 90 m (295 feet) deep near the rim, though most grooves average 100–200 m (328–656 feet) wide and 10–20 m (33–65 feet) deep. Many of the grooves have a pitted appearance resembling chains of craters. Remarkably, some of the pits appear to have slightly raised rims similar to those of impact craters. But the gravity field of Phobos is too small for these pits to be secondary impact craters from Stickney.

Instead, these raised rims could be regolith deposits blown into piles as gases escaped explosively through the grooves. It has been suggested that the large amounts of gases incorporated in the carbonaceous chondritic materials thought to be inside Phobos could have been liberated catastrophically by the shock and heat of the formation of Stickney.

Other scientists feel that the pits are collapse craters produced by sifting of the regolith into the fractures. They argue that the raised rims on the pits are an illusion produced by shading artifacts from

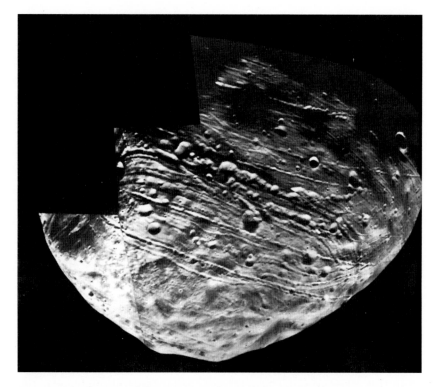

Figure 108. The rim of the crater Stickney *(left)* showing ejecta (the mottled hummocky region) and grooves. The grooves decrease in prominence away from the crater. Although these grooves appear to be associated with Stickney, their distribution is more consistent with the direction of Mars, suggesting that they may have been caused by tidal forces from Mars. (Courtesy NASA) (P-19133)

the type of image processing done on most images. The only way to resolve this debate is for additional detailed topographic information to be collected on these pits.

There are others grooves on Phobos that appear to be part of a global fracture system not related to Stickney. The grooves that lie radial to Stickney are thought to have cracked opened as a result of its formation, but the other grooves may be the result of strong tidal stresses exerted on Phobos by Mars. The close proximity of Phobos to Mars results in strong tidal forces that constantly tug at the satellite, producing stresses that act to pull it apart.

Deimos

Deimos, like Phobos, is a small hunk of rock measuring about 25 by 12 by 10 km (15 by 7.2 by 6 miles). It is relatively smooth (Figure 109) and has a shape that is distinctively nonellipsoidal and irregular. Most landforms (mainly craters) on Deimos have a more subdued appearance than similar landforms on Phobos, probably caused by a thick blanket of regolith.

Why Deimos has a comparatively thick regolith compared with Phobos is a mystery. Regoliths on small bodies such as Deimos are generated mainly by the grinding up of the surface by impacts. Calculations of the generation of regolith by impact craters predict that because of its greater surface area Phobos has collected more craters, and its regolith should be ten times greater than that of Deimos. So why does Deimos have such a thick regolith?

Peter Thomas has offered a solution to this dilemma. He thinks that the large, irregular depression in the south polar region of Deimos may be the remnants of an ancient 11-km (6.6-mile) impact crater. This would be huge in relation to the size of Deimos. Thomas noted that because large craters excavate large volumes of rock, they produce large volumes of regolith. A large crater such as that proposed by Thomas could have been capable of generating most of the observed regolith blanket.

As with the crater Stickney on Phobos, Thomas's proposed crater would have nearly blown the satellite apart and produced a system of grooves. Instead, there is no global groove system, like that found on Phobos, radiating from this proposed crater. Perhaps the grooves on Phobos were formed from other causes, such as tidal forces, or perhaps the depression is not really a crater. Alternatively, formation of the crater Stickney could have produced shaking of the moon severe enough that most of the regolith was tossed into space and never recovered by Phobos.

The largest confirmed crater on Deimos is 2.3 km (1.4 miles) in diameter. There is a continuum of sizes of craters smaller than this crater. The size-frequency distribution of the larger craters is similar to that of craters on Phobos, but small craters are much less abundant on Deimos than on Phobos. The population of larger craters probably records the flux of impacting materials over the history of these moons and suggests that both moons are ancient; the popula-

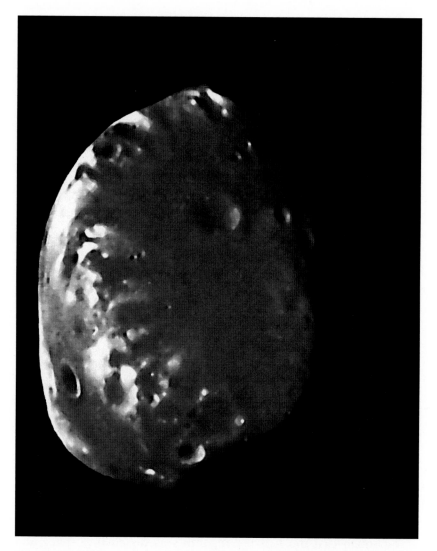

Figure 109. Viking took this picture of Deimos from a distance of about 948 km (570 miles). A ridge, seen on the left, separates two regions of the satellite. Bright patches occur along the ridge and bright streaks trail away from it. These bright streaks appear to be caused by regolith sliding or slumping down slopes. Such slides and slumps are also found on asteroids. (Courtesy NASA, Viking) (428B22)

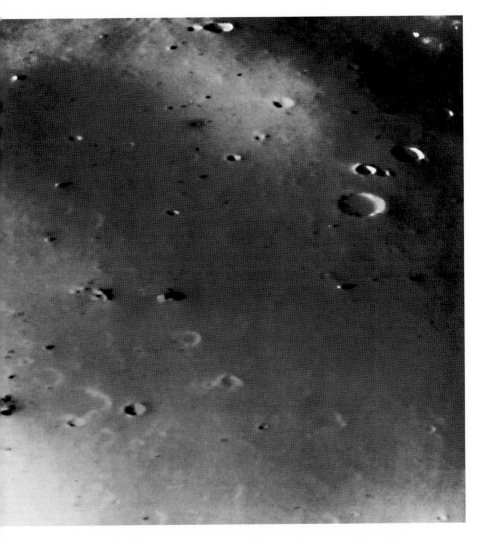

Figure 110. Filled craters and clumps of blocks, some up to 20 m (66 feet) across, litter the surface of Deimos near its South Pole. The picture was taken by Viking at a distance of 45 km (27 miles) from Deimos. (Courtesy NASA, Viking) (423A63)

tion of small craters appears to be reduced because of burial or mass wasting associated with the thick regolith.

Fine-grained material has ponded in the bottoms of all but the youngest craters (Figure 110). In some cases, this material is so thick that it forms flat floors in the impact craters. In other cases this material nearly fills the crater. It is probably the cause for the scarcity of small impact craters; although they are there this material has filled them.

Bright streaks extend down the slopes of some craters on Deimos. These streaks are probably fine-grained regolith that has slid or slumped downslope (Figure 109). On large bodies with greater gravity, such as Earth, this type of downslope movement produces landslides and is common. On tiny Deimos, the gravity field is too small to be the main cause of these slides and slumps. Instead, the pounding from the constant bombardment by micrometeoroids and the small but constant energy from solar radiation and solar wind may be enough to cause the formation of these streaks. These processes have been shown to be important factors in downslope movement of materials on our Moon and some asteroids.

As on the surface of Phobos, numerous blocks 20–30 m (66–98 feet) in diameter are found scattered around the surface of Deimos (Figure 110). These building-sized boulders are thought to be large chunks of impact crater ejecta that failed to completely escape the influence of Deimos. In contrast to those on Phobos, these blocks are widely distributed and difficult to relate to their source crater. Such a wider dispersion of ejecta on Deimos compared with that on Phobos may be an indicator that Deimos is composed of stronger materials than Phobos. Typically, compared with ejecta on bodies composed of weaker materials, ejecta may be dispersed at higher velocities and for greater distances on bodies composed of strong materials.

Chapter 9

Dreams and Schemes

NASA is pressing ahead with its intense program of the exploration of Mars. They plan to carry out this exploration with a fleet of robotic probes, each providing an increment of knowledge about Mars that systematically builds our understanding of whether life exists there or not. Other countries have joined this great venture. Sometimes new members of the Mars explorer club plan to go it alone and other times in strong collaboration with the United States. With this picture in mind, this final chapter focuses on the future exploration of Mars—when do we go, why do we go, how do we get there, and what do we do once we have arrived?

For any country that seeks to explore Mars, the answers to these questions are tied closely to national political objectives and, as a result, change with the times. For this reason, it should be kept in mind that frequently the dreams and plans to explore Mars have gone nowhere or floundered in midstream, even when they are the dreams of individuals at the highest level in U.S. government. For example, in 1991 President George H. W. Bush announced that the United States would aggressively pursue landing humans on Mars within a decade or two. This dream clearly did not happen but, just the same, showed vision and intent. Fortunately, there have been other dreams that have become realities.

The United States Goes to Mars

Currently the exploration of Mars is a major part of NASA's efforts in deep space exploration. What drives this focus on exploring Mars more than on any other planet? The answer has historical, mythological, and scientific roots, but the single thread that ties all of these together is our intense fascination with the possibility that Mars harbors life.

More than any other planet in the solar system, Mars has the necessary ingredients for life to have started and flourished. From our terrestrial experience with what it took to create life on Earth, early Mars appears to have had the necessary ingredients to also be a cradle of life. From our previous visits to Mars, we know that there are many places (most are underground) on Mars that currently have Earth-like environments and could provide comfortable homes for Martian organisms to thrive. Added to this, the more we learn about Mars, the more it appears to be a place that we could live some day—as Martian colonists.

Before we can think seriously about becoming Martian colonists, we must know enough about Mars to survive there. In the next section the strategy that the United States has developed for exploring Mars, with the intent of eventually having this knowledge, is laid out. The strategy dictates the methods (orbiters, landers, flyers, rovers, and so forth) to be used to explore as well as when and how.

The Strategy for Exploration

The exploration of Mars is not done in a random way. It is based on well-thought-out goals and strategies for systematically achieving those goals. Before NASA embarked on its long-term program of exploration of the solar system, it sought out the wise council of the National Academy of Science's Space Science Board on how to approach this endeavor.

In 1974–1975 the Space Science Board laid out its recommendations. In their answer they outlined a simple, yet profound methodology that continues to be a cornerstone of how NASA approaches exploration. They made these observations:

The investigation of any solar system object can be divided into three categories: reconnaissance, exploration, and intensive study. As a first step of our qualitative scale of progress on a given planet, we may speak of reconnaissance in which major characteristics are first sought and identified. Reconnaissance tells us qualitatively what the planet is like and provides enough information about the character of the planet and its environment to allow us to proceed to the stage of exploration of the planet. Exploration seeks to systematically discovery an understanding of the processes, history, and evolution of a planet on a global scale. In the final step, that of intensive study, sharply formulated specific problems of high importance are pursued in depth.

They outlined what type of mission fell into each of these categories:

Each of these phases are carried-out by missions that progressively increase in complexity (and cost). They begin with flybys designed to define and identify important characteristics of the planet in brief visits. These flybys fulfill the role of reconnaissance. These flybys should be followed by orbiters and simple landers that are capable of staying near or on the target longer. They are used in the beginning of the exploratory stage. These spacecraft should carry experiments for specific investigations based on information obtained from the flyby missions and updated earth-based observations. These spacecraft should be instrumented to measure the planetary properties on a global scale by remote sensing. Another type of follow-on mission involves in situ measurements on the target object by means of entry probes (either atmospheric or hard landers). These missions are of great importance in defining the chemical composition of a planet. In the case of rocky planets (like Mars), these may be followed by soft landers. Which carry out a wide variety of geophysical, chemical, and biological investigations at the surface on the place of landing. The latter mission would represent more advanced systems and should, in general, follow after the orbital and entry-probe missions. The ultimate exploration technique utilized or contemplated at present involves the return of samples from the surface for examination in terrestrial laboratories.

Even at the time of their report and on the threshold of the Viking landing, it was clear that the exploration of Mars was well into the exploration phase. They observed that "It is apparent that the reconnaissance mode of study of the inner solar system is completed and that a new generation of exploratory studies is just beginning." Nearly a decade and a half after Viking, the United States took the next logical step in Mars exploration: Mars Observer. This step ended in disaster. The Mars Program was in crisis. How would NASA respond?

NASA's Mars Exploration Program

In 1993, out of the ashes of Mars Observer, NASA's administrator Daniel Goldin took a bold and visionary step. Instead of taking actions that would halt the exploration of Mars, as some feared, he pushed Mars to the top of NASA's priority list. This started what has been called the second great era of the exploration of Mars.

In a plan hatched between Goldin and Wesley Huntress, the NASA associate administrator for space science (who was in charge of all NASA's Mars missions), NASA would carry out a long-term, low-cost program for the exploration of Mars. It had three goals: searching for life, characterizing Martian geology, and working out the intriguing history of the Martian climate. They recognized that the thread that ran through these goals and related them to one another was water—the universal solvent. But the cornerstone, the main reason for going, was the search for life.

To garner the political, scientific, and technical backing to make this vision a reality took considerable skill. NASA had to convince several different critical sectors that aggressively exploring Mars was a winning idea. The parts of the government that control budgets and directions, the science community that would help set intellectual foundations and science goals, and the aerospace industry that would build the robots to carry out this vision were all systematically convinced. The results of this effort gave birth to the Mars Surveyor Program. This program is now called simply the Mars Exploration Program. Although the details of this program have changed since its beginning, its overall vision has stayed the same.

NASA's Mars Exploration Program, as originally envisioned by Goldin and Huntress, included two Mars vehicle launches every 26 months. Mars missions can be launched every 26 months during the episodic alignment of Earth and Mars (called launch opportunities), which facilitates using the minimum amount of fuel to make the long trip. The plan was that at least one of these launches would carry a lander. The other would be an orbiter that carried out science observations and acted as a communications link for the landers.

The hallmark of this program was its low cost and resilience. In a long-term program, many approaches to reduce costs makes sense as opposed to a "one shot" program. For example, technologies developed for one mission can be used on the later missions and management teams can be shared between projects to make each mission cheaper. A long-term program also has the advantage that it can be done in a series of smaller steps, instead of one giant expensive one.

This new long-term program for Mars Exploration began with the launch of Mars Global Surveyor in 1996 (see chapter 3). The mission duplicated most of the Mars Observer science payload. Its companion mission, Mars Pathfinder, started its development as a technology demonstration mission and was later incorporated into the Mars Exploration Program. These missions were great successes, far exceeding expectations. Events took a turn for the worse during the next launch opportunity in 1998. This phase of the Mars Exploration Program ended in complete failure with the crash of all four U.S. Mars spacecraft. From the beginning of this program, NASA understood that the exploration of Mars, at best, was extremely difficult. Failures are to be expected, and historically half of the spacecraft sent to Mars have failed to accomplish their missions.

The failures of 1998 caused NASA to take a long, hard look at the approaches it employed to manage the Mars Exploration Program. They found poor management practices and serious overcommitment of resources; as a result the missions were developed with an unacceptably high level of risk. It was found that the Mars Surveyor Program launches in 1998 had been a disaster waiting to happen. NASA learned this bitter lesson quickly and set about redesigning its Mars Exploration Program. This time risk would be managed.

The next launch opportunity for Mars missions was in 2001. The Mars Surveyor Program 2001 mission was in its development phase when Mars Climate Orbiter and Mars Polar Lander crashed. As with the 1998 mission, the 2001 mission was to be a pair of spacecraft. The Mars Surveyor 2001 lander, a copy of the Mars Polar Lander, was to land in the equatorial highlands to explore ancient highland terrain. This too was not to be. Similarities in design of the Mars Polar Lander and the 2001 lander caused NASA to cancel the mission. After careful evaluation, the orbiter, Mars Odyssey, was completed and launched on schedule.

With the launch of Odyssey, a new phase of Mars exploration began, based on the lessons learned from the 1998 failures and the renewed commitment of NASA to the exploration of Mars. In announcing the new phase of Mars exploration in the fall of 2000, Scott Hubbard, director of the Mars Program at NASA Headquarters, Washington, D.C., said, "We have developed a campaign to explore Mars unparalleled in the history of space exploration. It will change and adapt over time in response to what we find with each mission.

It's meant to be a robust, flexible, long-term program that will give us the highest chances for success." He added, "We're moving from the early era of global mapping and limited surface exploration to a much more intensive approach. We will establish a sustained presence in orbit around Mars and on the surface with long-duration exploration of some of the most scientifically promising and intriguing places on the planet." The underpinning of this program remains strongly science. James Garvin, NASA Mars Exploration lead scientist at Headquarters, said, "The scientific strategy developed for the new program is that of first seeking the most compelling places from above, before moving to the surface to investigate Mars."

Following the Mars Odyssey orbiter mission, NASA plans to return to the surface of Mars in 2003 with the twin Mars Exploration Rovers (Figure 111). These rovers are planned to land in a manner similar to Mars Pathfinder, but with a modified system of air bags that allows a large spacecraft inside. The lessons learned from Sojourner will be used to help larger rovers roam the surface more easily. These rovers will land in the equatorial region to begin further characterization of the surface and search for signs of life. Of the possible landing sites, one area in particular has caught the attention of scientists. Measurements by the Mars Global Surveyor's thermal emisson spectrometer indicate that the rocks in the Terra Meridiani region contain a high concentration of the iron-bearing mineral hematite. Hematite is a mineral commonly formed in water-rich environments on Earth, such as warm springs. Because warm springs are a prime site to expect life, the Terra Meridiani region is a leading candidate as a landing site for one of the Mars Exploration Rovers.

NASA plans to launch a powerful scientific orbiter in 2005. The mission, called the Mars Reconnaissance Orbiter, will focus on analyzing the surface at ultra-high resolution in an effort to follow the tantalizing hints of water from the Mars Global Surveyor images and to bridge the gap between surface observations and measurements from orbit. The Mars Reconnaissance Orbiter will measure the Martian landscapes at 20- to 30-cm (8- to 12-inch) resolution, enough to observe rocks the size of beach balls.

Added to these larger missions, NASA envisions that the low-budget "Scout" mission might become an important part of this new program. Scouts will have budgets about the size of that of Mars Pathfinder and address goals of relatively narrow scope. These small

Figure 111. Two Mars Exploration Rovers are being developed for a 2003 launch to Mars. (Courtesy NASA)

missions might involve airborne vehicles (e.g., balloons and airplanes) or small landers, as an investigation platform. Exciting new discoveries could come from this small-mission approach either through the airborne scale of observation or by increasing the number of sites visited. The first Scout mission launch may be as early as 2007.

Though detailed plans for the Mars Exploration Program beyond 2009 are purposefully vague to preserve flexibility, NASA is hoping to launch its first sample collection mission early in the second decade of 2000. But Mars sample collection missions are costly. It will be, by far, the most complex robotic mission ever flown. There is no way around this and by some estimates the cost could be 3–4 billion dollars. The designers at Jet Propulsion Laboratory have done their best to devise methods to get samples off the surface of Mars and to Earth with relative economy (relative is the operative word here).

One of these concepts uses a small, two-stage, solid-propellant rocket carried to the surface of Mars to lift the samples to orbit (Figure 112). This small rockets owes its heritage to the rockets developed by the military in the 1950s that took very small payloads to Earth's orbit from the battlefield. In this concept, Mars samples

Figure 112. One option for a Mars sample collection mission is to collect samples and ascend to a rendezvous in orbit with a sample return orbiter. A small Mars ascent rocket is shown here lifting the samples to orbit. (Courtesy NASA)

would be collected, packed into the small rocket, and launched to Mars orbit. Once in Mars orbit, the small rocket would rendezvous with a "mother ship," transfer the samples, and the mother ship would carry the samples to Earth. An entry capsule carried by the mother ship would separate and make a direct entry and land. This method is thought to be the best option for preventing biological contamination of Earth (a rendezvous with a space station or orbital capture by a shuttle represents a greater risk).

Once samples are on Earth, they will require thorough examination to ensure that they contain no hazardous organisms and are safe to expose to Earth's environment (and for them to be exposed to Earth's environment). As with the Apollo lunar samples, they must be carefully curated after this examination and kept in a place where they will be preserved and distributed to the science community for study. For Mars samples, this place will be the Astromaterials Curatorial Facility at the Johnson Space Center.

Selection of the mission after 2007 must be made carefully. Missions to Mars are expensive. It takes 3–4 years to develop and fly a simple mission. A complex mission takes even longer. But on the other hand, because mission selection is partly dependent on what has been learned about Mars from previous missions, selection of specific options should not be done too far in advance. This new knowledge base is the foundation for developing the next series of major questions that have grown from that knowledge. A flexible approach allows new technologies to be used as they are developed and the United States to incorporate the plans of other countries for Mars exploration.

Europe Goes to Mars

NASA is not alone in its plans to explore Mars. The Europeans are also aggressively pursuing their own Mars program. Mars missions are planned by the European Space Agency as well as by individual European countries. These missions are carefully planned in collaboration with NASA to accomplish goals that are complementary with those of U.S. missions and when possible to provide communications support for each other's landers and rovers.

The European Space Agency will launch Mars Express in 2003 (Figure 113). Mars Express is an orbiter designed to map the distribution of water on Mars, measure the thickness of polar deposits,

Figure 113. Mars Express is the European Space Agency's first Mars mission. (Courtesy European Space Agency)

probe the ionosphere of Mars to study the interaction of the atmosphere and solar wind, and measure the physical properties of the surface. Mars Express may also carry a small lander, called Beagle 2, that is scheduled to land in the Isidis region.

It is clear that a Mars sample collection mission is expensive and that it will be difficult to fit into the low-cost mold of the U.S. Mars Exploration Program. The Europeans are eager to be partners in this great adventure and have begun to define roles for themselves. In particular, France and Italy are most interested and are planning to provide major components. For example, the Italians are planning to build the G. Marconi communication orbiter for launch in 2007 to Mars. This mission could handle the critical communication job associated with returning a sample from Mars. The French are also considering a major role. In particular, they may provide the "mother ship" that snatches the sample from Mars orbit and brings it to Earth. A test of this spacecraft is being considered in 2007, and it may deploy a network of four small landers, called NetLanders, to study the interior and atmosphere of Mars.

In the context of international exploration of Mars, it should be kept in mind that the Japanese and Russians are capable players. The Japanese Nozomi mission is a major start of Mars exploration by Japan, and although the Russian planetary program is currently stalled, it is capable of mounting a strong effort in Mars exploration.

With the missions from these countries joining NASA's armada of spacecraft preparing to go to Mars, it can truly be said that this second great era of Mars exploration has begun. Only time will tell how long this era will last and if it will lead directly to the next great adventure—human exploration of Mars.

The First Martians

Exploration of Mars by humans is the next logical step after robotic spacecraft build our understanding of the Martian environment and expand our search for life (Figure 114). There are many concepts for how this can be done. All face the same problems: mission complexity and the long duration. To begin with, we face the same launch opportunity spacing and flight time duration as robotic missions. Launch opportunities open only every 26 months and the journey will take between 4 and 6 months, one way. The flight time can be shortened using advanced propulsion systems, but it will still take several months.

In planning the first human mission, the 26-month intervals for launch opportunities must be considered. This, together with the duration of the flight to Mars, strongly influences the type of missions that are possible and the length of time on the surface before returning. Because of these constraints there are two choices for returning. After arriving on Mars, astronauts must either begin their return home after only several days or stay there approximately 2 years until the next opportunity to return opens. A mission with a short stay on the surface would require astronauts to carry all their supplies with them. A mission with a long stay on the surface would require astronauts to partially live off the land.

Mission planners favor long stays on the surface. More time on

Figure 114. Humans will eventually explore Mars, and at some point in the future they will come to stay—but when? (Courtesy NASA)

the surface affords more time to explore and to begin building the infrastructure for later, longer-duration stays. One such scenario being studied by NASA for our first manned mission to Mars uses as its key element the building of a critical infrastructure before humans are sent. A cargo vehicle that delivers supplies to the surface of Mars, a rocket to get from the surface up to Mars orbit, and an Earth-return vehicle would be pieced together in Earth orbit and launched to Mars. Once these are in place, possibly at the next launch opportunity, astronauts would follow. When they arrive on Mars they would land near their cargo ship and ascent rocket and begin to set up their outpost.

An important part of this mission scenario is that the astronauts would build small factories to manufacture some of what they need to live, such as food, fuel, and oxygen. Like the Apollo astronauts, they would also conduct a wide assortment of scientific experiments. Searching for life (extant or extinct) would be one of the main objectives.

At the end of their stay, the astronauts would pack up the cargo of samples and launch to orbit using the ascent vehicle that had been delivered to the surface before the crew arrived. The ascent vehicle would be fueled by propellant they had manufactured on Mars.

Once in orbit the astronauts and their cargo would transfer to the orbiting Earth-return vehicle and start their journey home. The entire voyage to Mars and back would have lasted nearly 3 years.

This first step would soon be followed by more missions that carry more equipment and crews to build larger and more permanent bases. Eventually, these bases (or base) would grow to become self-supporting and large enough to sustain a permanent human presence on Mars—the first human outpost on another planet. The outpost would grow and others would spring up. The next natural step is very difficult—colonization of Mars. Granted, this sounds like a grandiose plan, but it is indeed a natural progression. The timetable is the only thing in question.

Colonization and Terraforming

Our ability to colonize Mars is severely limited by the harsh Martian surface environment. It is deadly to humans that lack expensive life support systems. Its surface is cold and dry, has little air and nothing to eat, and is bombarded by deadly radiation from the Sun. How can these obstacles be overcome to allow humans to survive for long periods on Mars?

One intriguing concept has been advanced: terraforming. Martian terraforming envisions human-engineered changes of the Martian environment that transform its surface into an Earth-like environment. Central to terraforming, according to Robert Zubrin of Pioneer Astronautic and Chris McKay of NASA's Ames Research Center, is the belief "that sufficient CO_2 to form 300 to 600 mbars atmosphere may still exist in volatile form, either absorbed in to the regolith or frozen out at the south pole. The CO_2 may be released by planetary warming, and as the CO_2 atmosphere thickens, positive feedback is produced which accelerates the warming (primarily through greenhouse heating). Thus it is conceivable that by taking advantage of the positive feedback inherent in Mars' atmosphere/regolith CO_2 system, that engineering can produce drastic changes in climate and pressure on a planetary scale." How can this be done?

The main ingredient in most terraforming schemes involves the initial heating of the surface to drive volatiles trapped in the subsurface into the atmosphere. Most of these methods sound like schemes right out of science fiction movies. Zubrin and McKay proposed several ways to produce enough heat. They suggested that huge mirrors "with dimensions on the order of 100 km radius" could be built to orbit Mars, sending beams of focused light to strategic sites to melt reservoirs of volatiles. Alternatively, asteroids from the outer solar system are thought to contain ample ammonia, a powerful greenhouse gas. Steering an asteroid's elliptical orbit to intersect the orbit of Mars would eventually produce an impact that could deliver enough ammonia to start the Martian atmosphere on a path to global warming through greenhouse heating. They have also suggested another innovative idea that uses less-dramatic means and is a remarkable reversal of what Earthlings want for Earth: that the hydrocarbons produced as by-products of Martian industry could contribute substantially to greenhouse heating. Consequently, future Martian industry could help terraform the planet.

In a much slower variation of the concept of terraforming, oxygen-producing algae could be introduced in the polar regions and at the seeps in the equatorial regions of Mars where water and water ice are available at the surface. As these rugged little algae grow, they would generate oxygen and water vapor as by-products. As these by-products are introduced into the atmosphere, it would slowly be transformed to a warmer, thicker atmosphere rich in oxygen and water vapor. This method is very slow and would require that we wait over a thousand years for results.

Whether a Blight or a Bearer?

The concept of terraforming is dramatic and would certainly alter the pristine nature of the Martian environment. The planet Mars as we know it would be forever changed. The climate and any possible indigenous life would feel the effects. It is clear that before we alter the surface environment of another planet, considerable thought should be given to the outcome.

But putting the concept of terraforming aside, what about unintended changes to the Martian environment brought on by the mere acts of our current exploration? Most will probably be biological. For example, microbes have already been delivered to the surface of Mars by our landers. Were these microbes able to survive the caustic soils and intense radiation to gain a foothold on Mars? Will they displace indigenous species or mutate into new forms and greet us as pathogens or be mistaken for Martian life? Though the answer to these questions is probably no, there is also a finite possibility that it could be yes. How will we know?

What about when humans show up on the Martian doorstep? By all accounts, no matter how well planned or prepared, the first visit by humans will mark the point where the pristine biologic environment of Mars will be dramatically changed. It is inescapable. Humans are complex biological organisms and along with their life support systems carry a wonderful assortment of microorganisms. As a result, it will be impossible to sterilize every footprint of each human to keep Mars pristine without first destroying all these organisms, including the number one carrier—humans. It is conceivable that we might become the carriers of a "plaguelike" epidemic to Mars, much as Ray Bradbury described the introduction of chicken pox in his 1950 book *The Martian Chronicles*.

It is conceivable that some day new Martians, colonists from Earth, as they stand by the reflective waters of a Martian canal, may say much the same thing as Bradbury's colonists:

"I've always wanted to see a Martian," said Michael. "Where are they, Dad? You promised."

"There they are," said Dad, and he shifted Michael on his shoulders and pointed straight down.

The Martians were there. Timothy began to shiver. The Martians were there—in the canal—reflected in the water. Timothy and Michael and Robert and Mom and Dad.

The Martians stared back at them for a long, long silent time from the rippling water. . . .

Glossary

Absolute age   The age of a geologic unit measured in years.

Accretion   The gradual assembly of a body from many smaller bodies; all planets and moons form by accretion.

Albedo   The reflectivity of an object; light and dark are high and low albedo, respectively.

Altitude   Vertical distance above the surface of an object.

Aluminum   A metal element abundant on Mars; atomic number 13, atomic weight 26.9, symbol Al.

Anastomosing   Pertaining to channels that branch and rejoin, forming a braided pattern.

Andesite   A dark gray, fine-grained, relatively silicon-rich volcanic rock.

Anomaly   Something markedly unexpected or a value different from many others in a data sequence.

Anorthosite   A slowly cooled rock made up almost solely of the calcium-rich minerals plagioclase and feldspar.

Aphelion   Point in its orbit at which a planet is farthest from the Sun.

Argon   A noble gas element produced by the Sun and by the natural decay of radioactive potassium; atomic number 18, atomic weight 39.9, symbol Ar.

Ash   Very small fragments of lava sprayed out of a vent, cooled quickly, and deposited as a blanket of debris.

Ash flow (pyroclastic flow)   Ash that behaves as a single massive fluid; may travel great distances.

Asteroid   A small body, usually less than a few hundred kilometers in size, that orbits the Sun as an independent planet.

Asthenosphere   The partially melted or weak layer underlying the outer rigid lithospheric layer of a planet or satellite.

Astronomical   Dealing with objects and phenomena in the sky; their observation, and the instruments used to observe them.

Atmosphere   The envelope of gas that commonly surrounds planets.

Atom   The smallest particle of an element that retains the properties characteristic of that element.

Atomic number   The number of protons in an atomic nucleus.

Atomic weight   The total number of protons and neutrons in a nucleus.

Axis   Imaginary line about which a planet or satellite rotates; intersects the surface at the poles.

Ballistic trajectory   The looping path that an ejected particle travels while in flight.

Basalt   A dark, fine-grained volcanic rock rich in iron and magnesium.

Basin   A very large impact crater, usually greater than 300 km (180 miles) in diameter.

Bedrock   The intact layer of rock below the regolith.

Bombardment   The collision of a planet with asteroids, repeatedly over time.

Breccia   Rock and mineral fragments cemented together.

Bulk composition   The chemical and mineral composition of an entire planet or satellite; it cannot be measured directly but must be calculated.

Butte   An isolated, flat-topped hill or small mountain with steep sides.

Calcium   A light metal element, abundant in the Martian crust; atomic number 20, atomic weight 40.1, symbol Ca.

Caldera   A large volcanic depression, often formed by collapse caused by the withdrawal of underlying magma.

Canyon   A steep-sided, deep valley.

Carbon   A light element, a major component of the Martian atmosphere; atomic number 6, atomic weight 12, symbol C.

Carbon dioxide   Gas made of one carbon atom and two oxygen atoms; formula CO_2; the main component of the Martian atmosphere.

Carbon monoxide   Gas made of one carbon atom and one oxygen atom; formula CO.

Cartography   The science and art of expressing graphically, by means of maps and charts, the physical features of a planet.

Chaos (pl., Chaos)   Latin, meaning boundless empty space; a jumble of crustal blocks in Mars collapsed terrain.

Chaotic terrain   Low region within the heavily cratered uplands that appears to consist of irregular, blocky, fractured terrain.

Chasma (pl., Chasmata)   Latin, meaning a large canyon or gorge; used on Mars as part of a feature name to indicate that it is a canyon.

Cinder cone   A hill produced by the buildup of ash or other pryoclastic fragments around a volcanic vent.

Cirque   A deep, semicircular, bowl-shaped erosional feature formed by a glacier and located at the head of a glacial valley.

Clast   A fragment of rock or mineral in a sedimentary rock.

Comet   A small body of icy and dusty matter that revolves around the Sun.

Compression   Forcing mass into a smaller volume (increases density).

Conduction   The transfer of heat from a hotter region to a cooler one by the vibration of atoms.

Conglomerate   A rock made up of other rocks; distinguished from a breccia by the rounded clasts.

Convection   The process of transferring heat by a fluidlike motion of material driven by density variation of the material.

Core   The central zone of a planet, usually made up largely of metallic iron.

Crater   A circular depression on a planet.

Creep   Slow, forward movement of grains produced by the force of gravity or drag from the wind or water.

Crust   The outer zone of a planet, made up of relatively low-density rocks; this layer is chemically distinct from the underlying mantle.

Crustal dichotomy   Strong hemispheric contrast in the physical nature of a planet's crust.

Crystal   A substance possessing permanent and regular internal order of structure.

Deflation   Removal of loose, granular particles by the wind, caused by fewer materials coming upwind than leaving downwind.

Density   The ratio of an object's mass to its volume.

Differentiation   A separation or segregation of different kinds of materials in different layers in the interior of a planet.

Dike   A blade-shaped volcanic body caused by injection of magma along a subsurface fracture.

Dome   A bulbous, hill-like volcano.

Duricrust   Hardened crust of soil in semiarid climates, formed by precipitation of aluminous, siliceous, and/or calcareous salts at the surface as groundwater evaporates.

Dust   Material of silt (1/16–1/256 mm) and clay (>1/256 mm) size blown by the wind.

Dust devil   Swirling, vertical updraft of air developed by local heating of the air above the flat desert floor.

Dynamo   Relative motion of electrically conducting materials that are capable of generating a magnetic field, such as the solid metallic inner core and the molten outer core of a planet.

Eccentricity   The ratio of the distance between the foci of an ellipse and the major (or longest) axis.

Ecliptic plane   The plane of a planet's orbit around the Sun.

Ejecta   Material excavated during the formation of an impact crater and deposited around the crater.

Ejecta blanket   Debris resulting from an impact that surrounds the crater rim.

Electron   A negatively charged subatomic particle that normally moves about the nucleus of an atom.

Element   A substance defined by number of electrons and protons, having definite and distinct chemical and physical properties.

Elliptical orbit   A noncircular orbit shaped like an ellipse, having a low and high altitude.

Endogenic   Process or features resulting from processes originating within the body (e.g., volcanism or tectonism).

Equilibrium   State in which all conditions remain constant with respect to each other, although the set as a whole may change.

Erosion   The destruction of planetary surface features by some process.

Escape velocity   The speed an object must achieve to break away gravitationally from another body.

Exosphere   The outermost region of the atmosphere that grades continuously into space.

Fault scarp   A cliff formed by the surface expression of a fracture in a planet's crust.

Feldspar   A rock-forming mineral that is rich in aluminum and may contain calcium, sodium, and potassium.

Flood basalt   Basalt that forms thick, extensive volcanic plains.

Flow front   The terminal end of a lava flow.

Flyby   A mission technique in which a planet is examined by a spacecraft as it flies by the planet on its way to another destination.

Fracture   A zone of failure in a geological material.

Gamma radiation   High-energy radiation produced during a nuclear reaction.

Gamma-ray spectrometer   An instrument that measures gamma radiation as a function of energy; can measure chemical composition of materials and planetary surfaces.

Geological history   The natural evolution and history of a planet, reconstructed through study of its rocks and surface features.

Glass   A natural material that possess no internal order; a liquid of high viscosity.

Graben   Two parallel, inward-dipping faults with a down-dropped block between them.

Granite   A course-grained igneous rock containing mainly quartz, potassium, and aluminum silicates.

Gravity   The tendency of matter to attract itself.

Gravity map   A map showing the variation in gravitational attraction across the surface of a planet, caused by variations in the internal density of the planet.

Hydrogen   The lightest and most abundant element in the universe; atomic number 1, atomic weight 1, symbol H; main component of the solar wind.

Hydrous phase   A substance containing water.

Hypothesis   An assumption or idea that is tested by experiment or observation.

Igneous rock   A rock crystallized from a liquid (magma).

Impact Geological process resulting from the collision of objects in space.

Impact melt Target material that was melted by the heat generated by an impact.

Inclination (of an orbit) The angle between the orbital plane of a body and the ecliptic plane.

Infrared Light having a wavelength greater than that of red light; invisible but felt as heat.

Iodine A nonmetallic element; atomic number 53, atomic weight 126, symbol I.

Ion An atom that has become electrically charged by the addition or loss of one or more electrons.

Ionosphere The upper region of an atmosphere in which many of the atoms are ionized.

Iron A common metal and rock-forming element; atomic number 26, atomic weight 55.8, symbol Fe.

Isotope A variant of an element; caused by an excess or deficiency of neutrons.

Kinetic energy Energy associated with motion; the kinetic energy of a body is one-half the product of its mass and the square of its velocity.

Krypton An inert rare gaseous element; atomic number 36, atomic weight 83.8, symbol Kr.

Labyrinthus Latin, meaning labyrinth; used to describe a complex of intersecting valleys or canyons on Mars.

Laser altimetry The process in which the topography of a planet can be determined by measuring the time it takes laser pulses to travel round trip between a spacecraft and the planet's surface.

Latitude Angular distance north and south from the equator of a planet, measured in degrees.

Launch window A range of dates during which a space vehicle can be launched for a specific mission without exceeding the fuel capabilities of the system.

Lava Liquid rock extruded onto a planetary surface.

Lead A heavy metal element; atomic number 82, atomic weight 207.2, symbol Pb.

Lithosphere The outer rigid layer of a planet or satellite.

Longitude Angular distance due east or west from the prime meridian of a planet.

Magma Liquid rock within the interior of a planet.

Magnesium A light metal element; atomic number 12, atomic weight 24.3, symbol Mg.

Magnetic anomaly A zone of a planetary surface where the magnetic field is either stronger or weaker than expected; any zone of intense magnetization on Mars.

Magnetic field A region in which a magnetic force is detectable everywhere.

Magnetized Material that is surrounded by lines of magnetic force; it may attain this magnetism by passing through an existing field or cooling in the presence of such a field.

Magnetosphere A region around a planet occupied by its magnetic field.

Mantle The middle layer of a planet, between the crust and the core, and chemically distinct from them.

Mascon Acronym for "mass concentration"; a zone of anomalously high density within Mars, detected over circular basins; first found on the Moon.

Mass A measure of the total amount of material in a body.

Mesosphere A zone of the atmosphere above the stratosphere where the temperature remains nearly the same with increasing altitude.

Meteor A rock or ice particle entering Earth's atmosphere at high velocity, forming a tail of ionized gas.

Meteorite A meteoroid that strikes the surface of a planet.

Meteoroid A meteoritic particle in space before any encounter with a planet.

Mineral A naturally occurring substance having a crystalline structure; rocks are aggregates of minerals.

Molecule A combination of two or more atoms bound together; the smallest particle of a chemical compound or substance that exhibits the chemical properties of that substance.

Mons (pl., Montes) Latin, meaning a mountain; used in names of features on Mars to indicate that they are mountains.

Morphology The shape and structure of a landform.

Multiring basin A large feature of impact origin having multiple, concentric-ringed structures.

NASA National Aeronautics and Space Administration, America's space agency.

Nebula Large cloud in space in which new stars and planets are created.

Neon A noble gas element; atomic number 10, atomic weight 20.2, symbol Ne.

Neutron A subatomic particle with no charge and with mass approximately equal to that of a proton.

Nickel A heavy metal element; atomic number 28, atomic weight 58.7, symbol Ni.

Nitrogen A light gas element; atomic number 7, atomic weight 14, symbol N.

Noble gas Rare and generally inert gases such as helium, neon, radon, etc.

Nucleus The heavy part of an atom, composed mostly of protons and neutrons, about which the electrons revolve.

Obliquity The tilt of a planet's rotation axis from perpendicular relative to its orbit.

Occultation An eclipse of a star, planet, or spacecraft by a satellite or planet.

Opposition The position of a planet when it is on the opposite side of the Earth from the Sun.

Orbit The path of one object around another.

Outcrop A natural exposure of bedrock.

Oxygen A common and abundant colorless gaseous element; atomic number 8, atomic weight 16, symbol O.

Patera (pl., Paterae) Latin, meaning a shallow dish or saucer from which a libation was poured; used on Mars to indicate that a feature is a low, broad mountain, generally volcanic in origin.

Payload The cargo or object of value carried into space by a booster rocket.

Perihelion The closest approach of a planet to the Sun.

Perturbation The deviation from the expected motion of a body caused by the gravitational influence of other bodies.

Photodissociation The process of molecules breaking apart when they absorb energy from sunlight.

Plagioclase A silicate mineral rich in aluminum and calcium or sodium; a subset of the feldspar group; common in anorthosite.

Planitia (pl., Planitiae) Latin, meaning plains; used on Mars to indicate that a region is composed of plains.

Planum (pl., Plana) Latin, meaning even, flat; used on Mars to indicate that a plains area is exceptionally flat.

Plasma A hot, ionized gas.

Plate tectonics Motion of a planet's lithosphere, causing fracturing of the surface into plates.

Potassium A light metal element; atomic number 19, atomic weight 39.1, symbol K.

Precession A slow rotation of a planet's axis due to the gravitational effects of a larger body on the planet's equatorial bulge.

Prime meridian The meridian of zero degrees longitude on a planet.

Projectile An object moving at high speed due to the influence of an external force.

Proton A heavy subatomic particle that carries a positive charge; one of the two principal constituents of the atomic nucleus.

Province A region of similar terrain or a collection of geological units with similar or related origins.

Pyroclastic Literally "fire-broken." Fragmental rocks, including ash, produced in explosive volcanic eruptions.

Pyroxene A magnesium- and iron-rich silicate mineral; an important constituent of igneous rock such as basalt.

Quartz A mineral of the compound silicon dioxide (SiO_2), very common on Earth.

Radiation The radiant transfer of energy through space.

Radiogenic isotope An element that naturally and spontaneously decays into another element.

Reflectance The amount of light reflected off a body or object in space.

Regolith The unconsolidated mass of debris that overlies bedrock on a planet.

Relative age The age of a feature or geologic unit in relation to other features or geologic units.

Resolution The capability of a system to make clear and distinct the separate components of an object.

Rille A linear or sinuous depression, commonly found in volcanic terrain.

Rings Concentric terrain elements that surround or lie within basins on the terrestrial planets.

Rock An aggregate of minerals; main constituent of the terrestrial planets.

Rotation The spinning of an object about a central axis.

Saltation Transport of sand by wind, in which individual particles continually impact, bouncing up or ejecting other particles.

Sand Particles 1/16–2 mm in diameter.

Sapping The process of undermining and removal of material by groundwater flow.

Sediment Loose material, formed by decomposition of rocks, that is transported from its place of origin and deposited elsewhere.

Sedimentary rock Rock formed from deposits of sediments cemented together.

Shield volcano A broad, low-relief, volcanic construct made up of flows of relatively fluid lava, usually basalt.

Shock melt Material melted by the passage of a shock wave.

Silicate A mineral group in which the silicon tetrahedron (a single silicon atom surrounded by four oxygen atoms, SiO_4) is always part of the structure; the largest mineral group, also called the "rock-forming" minerals.

Silicon A common light element; atomic number 14, atomic weight 28.1, symbol Si.

Slope Angle of tilt of a local surface.

Slope wind Wind induced by the thermal or mechanical effects of sloping terrain.

SNC meteorites Shergottites, nakhlites, and chassigny: a group of meteorites for which Mars is suggested as the parent body.

Soil Loose material that forms in place by weathering of rock material and that may contain zonal structure recording surface-atmosphere interactions; on Earth, the top weathered layer of terrestrial lithosphere that is exposed to atomspheric and biotic effects.

Sol One Martian solar day, equivalent to 24.66 Earth hours.

Solar nebula A cloud of gas and dust from which the solar system is presumed to have formed.

Solar wind A stream of electrically charged gas particles, mostly hydrogen, emanating from the Sun.

Spectral data Data dealing with the precise measurement of the color of planetary surfaces to determine composition.

Spectroscopy A measurement technique of determining energy intensity (reflected or emitted) as a function of wavelength.

Spectrum A plot of intensity as a function of wavelength.

Spin axis An imaginary line about which an object rotates.

Splotches Dark markings within Martian craters caused by removal or deposition of windblown material.

Stable isotope An isotope that is not radioactive.

Stratigraphy In planetary usage, the study of the relative age of rock units as determined by superposition and embayment relations and impact crater densities.

Striae A series of grooves etched in a rock surface, usually by flowing streams or ejecta flow and oriented parallel to the direction of movement.

Strontium A metal element; atomic number 38, atomic weight 87.6, symbol Sr.

Subcrustal erosion Removal of material from the base of the crust or lithosphere by motion of the underlying mantle.

Subduction The descent of a tectonic plate into the mantle or the downwelling of the lithosphere as a part of convective mantle circulation.

Subsolar The point on the surface of a planet where the Sun lies directly overhead.

Sulfur A light, nonmetallic element, atomic number 16, atomic weight 32.1, symbol S.

Summit pit A depression occurring at or near the top of a volcano.

Superposition The position of one feature on top of another feature or surface.

Suspension Mechanism of transport of fine material by the wind in which the movement of the wind keeps the particles aloft.

Talus Collection of debris that has fallen down a slope surface and formed a surface with lesser slope.

Target material The surface material into which an impact occurs.

Tectonism The process of deformation of planetary crusts and surfaces by large-scale forces associated with the dynamics of the mantle.

Tension   The condition of being pulled or stretched.

Terminator   The line between day and night side.

Terraforming   The changing of an alien landscape into one more suitable for human beings.

Terrestrial   Of or like Earth.

Terrestrial planet   A rocky planet, similar in properties to Earth; group includes Mercury, Venus, Earth, the Moon, and Mars.

Thermal history   The series of events throughout time caused by the internal temperature of a body.

Thermal tides   Atmospheric variations with a period of one solar day, induced by solar heating.

Thermokarst   Topographic depression with karstlike features (steep-sided pits) produced in permafrost regions and resulting from the melting of ground ice.

Thermosphere   The region in the upper atmosphere where the temperature rises rapidly with altitude; in the Martian atmosphere, this region occurs at 110–200 km (66–120 miles) altitude.

Tholus (pl., Tholi)   Latin, meaning a dome or cupola; used on Mars in the names of volcanic domes.

Thorium   A heavy metal element; atomic number 90, atomic weight 232, symbol Th.

Threshold wind speed   Value of wind speed at the initiation of grain motion.

Tide   The deformation of a body caused by the gravitational attraction of a second body.

Topographic datum   The reference surface from which elevations on a topographic map are measured; on Earth this datum is usually sea level.

Tributary canyon   Deep, local dendritic valley having a V-shaped cross profile and merging with the Valles Marineris or other troughs.

Trough   A deep, steep-sided depression; used for the Valles Marineris system of grabens.

Ultraviolet   Light having a wavelength beyond the violet end of the visible spectra.

Uranium   A heavy metal element, radioactive; atomic number 92, atomic weight 238, symbol U.

Vallis (pl., Valles)   Latin, meaning valley; used on Mars in the names of valleys.

Valley network   System of interconnecting drainages on the surface of Mars that may have carried fluids but lacks the bed form features that are direct indicators of fluid flow.

Vent   A hole in a planet from which volcanic products (lava, ash) may be erupted.

Vesicle A hole (preserved bubble) in a sample of lava rock; results from dissolved gas in the magma coming out of solution.

Viscosity A liquid's resistance to flow.

Volatile element An element with a relatively low boiling temperature.

Volcanic construct A hill or mountain made of lava and/or pyroclastic materials that are constructed around a volcanic vent.

Volcanic rock Rock formed by the cooling of liquid rock (lava) extruded onto a planetary surface.

Volcanism The planetary process of interior melting, the movement of liquid rock through a planet, and the eruption of such liquids onto the surface.

Water A compound of two hydrogen atoms and one oxygen atom; formula H_2O; it is a liquid at room temperature but readily freezes or vaporizes.

Wavelength The distance between wave crests; for visible light, the same as color.

Weathering Physical and chemical interactions between surface materials (including inertial volatiles) that lead to decomposition or alteration of rocks, minerals, or mineraloids and the possible formation of new phases (usually taken to mean at the scale of individual mineral grains).

Wind streak Light or dark patches emanating from topographic obstacles on the Martian surface and pointing in the direction that was downwind at the time of their formation; they can either be erosional or depositional, depending on conditions that existed at the time of their formation.

Wrinkle ridge A ridge or long, quasilinear topographic high on the surface of a planet due to compressional forces acting on the lithosphere.

Xenon A heavy, rare gaseous element; atomic number 54, atomic weight 131, symbol Xe.

Yardang Streamlined, aerodynamically shaped elongate hill oriented parallel to the direction of the wind.

Zenith The point directly overhead.

Further Reading

Beatty, J. Kelly, and Andrew Chaikin, eds. *The New Solar System*. 3d ed. Cambridge, Massachusetts: Sky Publishing Corporation, 1990.

Bergren, Laurence. *Voyage to Mars: NASA's Search for Life beyond Earth*. New York: Riverhead Books, 2000.

Caidin, Martin, and Jay Barbree. *Destination Mars: In Art, Myth, and Science*. New York: Penguin Putnam Incorporated, 1997.

Carr, Michael H. *The Surface of Mars*. New Haven: Yale University Press, 1981.

————. *Water on Mars*. Oxford: Oxford University Press, 1996.

Cooper, Henry S. F., Jr. *The Search for Life on Mars: Evolution of an Idea*. New York: Holt, Rinehart, and Winston, 1980.

Ezell, Edward C., and Linda N. Ezell. *On Mars: Exploration of the Red Planet, 1958–1978*. Washington, D.C.: NASA Special Publication 4212, 1984.

Glasstone, Samuel. *The Book of Mars*. Washington, D.C.: NASA Special Publication 179, 1968.

Greeley, Ronald. *Planetary Landscapes*. London: Allen and Unwin, 1985.

Horowitz, Norman H. *To Utopia and Back: The Search for Life in the Solar System*. New York: W. H. Freeman, 1986.

Hoyt, William Graves. *Lowell and Mars*. Tucson: University of Arizona Press, 1976.

Keiffer, Hugh, Robert Pepin, and Bruce Jakosky. *Mars*. Tucson: University of Arizona Press, 1992.

Nature. Insight Mars. *Nature* (London) 412:207–253, 2001.

Pritchett, Price, and Brian Miurhead. *The Mars Pathfinder Approach to "Faster-Better-Cheaper."* Dallas: Pritchett and Associates, 1998.

Raeburn, Paul. *Uncovering the Secrets of the Red Planet*. Washington, D.C.: National Geographic Society, 1998.

Sawyer, Kathy. *A Mars Never Dreamed of*. Washington, D.C.: National Geographic Society, 2001.

Sheehan, William. *The Planet Mars: A History of Observation and Discovery*. Tucson: University of Arizona Press, 1996.

Wilford, John Noble. *Mars Beckons*. New York: Alfred A. Knopf, 1990.

Index

Page numbers in italics indicate illustrations.